DEFENDING
YOUR CASTLE

Build Catapults, Crossbows, Moats, Bulletproof
Shields, and More Defensive Devices
to Fend Off the Invading Hordes

WILLIAM GURSTELLE

CHICAGO
REVIEW
PRESS

Copyright © 2014 by William Gurstelle
All rights reserved
Published by Chicago Review Press, Incorporated
814 North Franklin Street
Chicago, Illinois 60610
ISBN 978-1-61374-682-0

Questions? Comments? Visit www.DefendingYourCastle.com

Library of Congress Cataloging-in-Publication Data
Gurstelle, William.
    Defending your castle : build catapults, crossbows, moats, bulletproof shields,
and more defensive devices / William Gurstelle.
        pages cm
    Includes bibliographical references and index.
    ISBN 978-1-61374-682-0 (trade paper)
    1. Catapult—Design and construction. 2. Crossbows—Design and construction.
3. Fortification—Design and construction. 4. Siege warfare. I. Title. II.
Title: Build catapults, crossbows, moats, bulletproof shields, and more
defensive devices.

    U873.G87 2014
    623.4'41—dc23

                                                                    2013046712

Cover design and art: Damien Scogin
Interior design: Scott Rattray
Interior illustrations: Damien Scogin

Printed in the United States of America
5 4 3 2 1

My father, Harold, my brother, Steven, and my father-in-law, Milton, were in the US armed forces and served during World War II, the Vietnam War, and the Korean War, respectively. This book is dedicated to them and to all the veterans who have defended our castles.

Cave ne ante ullas catapultas ambules!

CONTENTS

CONTENTS

ACKNOWLEDGMENTS

I would like to express my gratitude to the people who saw me through this book; to all those who provided support, provided me a platform for talking things over, provided information and background, and offered comments, inspiration, and ideas. In particular, I thank my wonderful wife, Karen Hansen Gurstelle: creative, clearheaded, and thorough. This book could not have been written without her.

A CAST OF
BAD ACTORS

How far would you go to protect yourself, your family, and your stuff? Most likely, your answer is something along the lines of "well, as far as I have to." It's always been this way, since the first Neanderthal cave dweller barred access to his abode by pushing a large rock in front of the entrance.

There are two kinds of people in the world: those who have stuff and those who want to take other people's stuff.

This book describes methods, machines, and mechanisms for defending your home—your castle—from all sorts of bad actors: city sackers, pillaging pirates, bad guys, trespassers, brigands, hajduks, thuggees, dacoits, and every other organized, semi-organized, or anarchic band of evildoers seeking to bring mayhem and destruction to peaceful, law-abiding people like you.

After reading this book's title, you could be thinking that the projects and advice contained in this book are perhaps eight centuries out of date. Well, maybe they are. Certainly, the world is at least a bit more civilized now than it once was, and the threat of marauding hordes banging down your doors has lessened over time. But—and this is a big but—it has not gone away. Although we are currently experiencing the longest sustained peace between the largest and most powerful countries in history, things could always go haywire, and in a hurry. If that happens, few things will be as important as knowing how to mount a powerful and well-considered defense of your home and property. Protecting your castle is now, and always has been, a vitally important skill set.

There have been assaults on homes, villages, towns, and cities since the dawn of human civilization. Perhaps the earliest recorded siege is the destruction of the walled city of Jericho by Joshua and the Israelites in about 1350 BCE. Joshua's technique included blowing trumpets and marching around the walls while holding the Ark of the Covenant. Apparently this worked, because the walls, reads the Bible, did fall, allowing the Israelites to sweep into the city. The only Jerichonian to survive was a single woman to whom Joshua owed a debt of gratitude.

Such violence was all too common in ancient times. For example, diagrams and hieroglyphics on tomb walls in Egypt recount an ancient siege of the Hittite city of Dapur by King Ramses II in the 13th century BCE. Using scaling ladders, bows and arrows, and other implements of warfare particular to the earliest known wars, the Egyptian invaders were able to breach the walls and conquer the poor Hittites, causing them to exclaim, "We were crushed under your sandals and your might has penetrated our land!"

"Yes, but," you still may say, "that was then and this is now. I have no need to provide for my own defense. Why, I can call the police, or if worse comes to worst, the army will protect me." Perhaps you are correct. But again, perhaps not. On the night of April 24, 1184 BCE, the citizens of Troy slept soundly, feeling confident that their army and leaders would protect them from the Greek adventurers who stood outside the city walls. But at night, Athenian soldiers crawled out of the Trojan Horse, a large hollow statue that the city's leaders had stupidly brought inside the city walls, unaware of what and who were sitting silently inside. The spies opened the gates to the invaders and by morning, Troy had fallen.

In every century, groups, countries, and tribes of people have invaded other groups. Such invasions have been as dependable and as constant as death and taxes. In *Defending Your Castle*, ideas for protecting your homestead, school, or workplace against threats of many types are provided. There are two main themes to the book. The first covers projects and concepts related to real, material threats to your well-being. The concepts, ideas, and thinking processes detailed are suitable for use in a wide range of situations that I hope you never experience.

The second theme is more imaginative and probably a bit more fun. It attempts to answer a theoretical, yet still important, question. To wit, if you were attacked by a fierce, wild horde of Mongols, Vikings, Huns, Macedonians, or similar group, how could you protect yourself? It is (to me at least and since you bought this book, probably to you as well) an interesting question. To answer it we'll delve into the history of these peoples and explore their tactics with an eye to staying out of trouble. This section is a combination of science, hands-on DIY technology, history, and a special type of history called counterfactualism that seeks to explore what-if scenarios of our past.

1.1 Troy

The Procession of the Trojan Horse into Troy *by Giovanni Domenico Tiepolo, National Gallery, London*

THE FIRST COUNTERFACTUAL DIY BOOK, EVER

One of the oldest and yet most popular tropes in fiction is that of "alternative history" stories. Basically, such stories are based on speculation about what the world would be like if instead of x happening, y had happened. For example, how would the world be different if the Confederacy had won at Gettysburg, if Ottomans conquered Vienna in 1683, or if the "divine wind" had not blown Mongol invaders away from Japan's seacoast in 1281?

The earliest example of alternative history fiction I have found comes from the Roman writer Livy. His book *Ab Urbe Condita* (The city of Rome since it was founded) was written just after 9 BCE and is a monumental work, covering the history of Rome from its legendary founding by Romulus and Remus in 753 BCE to the death of Emperor Nero. For the most part, it is similar in form and tone to most classical Roman history works. Dense with dates and names, there are lots of accounts of the accomplishments the city's leaders and piles of impenetrable detail about heavy infantry battle tactics and cavalry charges. While history scholars may find it an interesting work, and certainly important, most readers would find it a dull slog indeed. There's little in it that's novel or terribly gripping. That is, until one reaches the ninth book in the series. For in book 9, paragraph 17, Livy does something no one had done before. He changes history, figuratively and literally.

After describing another in a long line of battles and sieges, Livy departs from his factual but dry retelling of the Second Samnite War and makes a sudden left turn into historiography's hall of fame: a writer goes "counterfactual" for what is likely the first time in written history.

(I) digress more than is necessary from the order of the narrative or by embellishing my work with a variety of topics to afford pleasant resting-places, as it were, for my readers

4

and mental relaxation for myself. The mention, however, of so great a king and commander [as Alexander the Great] induces me to lay before my readers some reflections which I have often made when I have proposed to myself the question, "What would have been the results for Rome if she had been engaged in war with Alexander?"

How different, indeed, would the world, 400 years after Alexander's army marched, have looked to Livy?

Quite a lot, apparently.

Livy wrote that he was certain that had Alexander marched west to Rome instead of east to Persia, Rome's legions would have triumphed over the Macedonians. Alexander's huge influence on the world, including the Hellenization of European culture, the rise of the Levantine Ptolemaic kings, and the chastening of Persian influence in the Mediterranean, would not have occurred. The world, surmises Livy, would be a completely different place. Perhaps he's right or perhaps he's not, but that's beside the point; Livy's sudden and novel what-if digression adds spark and interest to the page and makes his book one for the ages.

NEW COUNTERFACTUAL GENRES

Given a few facts and making use of the knowledge of human nature, all manner of thought-provoking alternative histories can be written. Since Livy's initial foray into counterfactual history writing, thousands, if not millions, of authors and thinkers have asked and speculatively answered similar questions. The popularity of such postulations is obvious now; they are clever techniques for probing the relative importance of a historical person or event.

For the most part, this is the stuff of novels and history textbooks. But more recently, counterfactualism is seen in modern types of media. Counterfactual and alternative history movies are popular and frequently successful. From *Back to the Future* to *Planet of the Apes* to *Groundhog Day*, counterfactualism is a popular fictive technique.

As you may expect, the Internet is filled with this as well. One well-known counterfactual meme kicked around the Internet asks: could a single US military battalion, with its modern firepower, communications, and training, defeat the entire Roman army? The question itself is interesting and easily lent itself to all sorts of juicy counterfactual analysis. So much so, that the idea became the basis for websites, books, endless comments on websites, and even a cable television show. *Defending Your Castle* explores the premise of a similar alternative history question: given modern knowledge of construction techniques and materials, could you successfully defend your home—your castle—from a horde of Hun, Viking, or Mongolian invaders?

The stories, histories, and projects that follow explore a variety of ideas for adding security to your castle (the word *castle* being a catch-all phrase to describe your house, your camp, your neighborhood, your school, or any place of refuge).

If you've paged through this book a bit already, you have likely noticed some fairly off-the-wall projects. For example, building a moat around your suburban home or adding observation towers to your backyard are provided as food for counter-historical thought—unless you're really distrustful of your neighbors! On the other hand, some projects are quite practical, depending on your needs and desires. Take a look at recent newspaper headlines, and I think you'll agree there are many people who just might benefit from a Kevlar backpack or hidden book safe. Many of the other projects reside in a gray area, straddling the not-so-well-defined border between useful, interesting, and fanciful.

1.2 A Well-Situated Castle *Fresco by Andrea Mantegna in the Palazzo Ducale di Mantova, Italy*

DEFENDING YOUR CASTLE

Let's conduct a thought experiment.

Take a moment and imagine looking out your front window and spotting a horde of men in front of your house, apartment, dormitory, or campsite. This group—a large, dirty, and noisy one—behaves in a manner that leaves little doubt that it is up to no good. A closer look through binoculars reveals the men are not particularly well equipped. The situation is confusing, worrisome, and a little baffling.

In this situation, as in most alternative history stories, the question of how this came to be doesn't really matter. Perhaps these people are remnants of a post-nuclear-war apocalyptic society, or they are the poor mutated remnants of a people scourged by a terrible, infectious disease, or maybe they are scared and confused Mongols accidently caught in a time warp and transported to the present time and place. Why they are here doesn't matter; what does matter is how you now choose to handle this situation.

The projects and historical vignettes in the chapters that follow are designed to provide the information and techniques you need to defend your castle and turn away modern-day thugs, brutes, hooligans, and other contemporary threats.

Ahead, you find many projects, each carefully designed and researched. There are myriad ways to accomplish an end, though, and there may be better methods and techniques for moat building or catapult construction than are provided here. What is here will go far to get you started, not just in defending your home, but also in learning to build interesting things and developing a new appreciation of history and science.

Genghis Khan, Alexander the Great, and Attila were all famous leaders who amassed giant empires. Theirs are household names. But although hardly known today, there were some defensive geniuses who stood up to and turned these invaders away. Flavius Aetius was a Roman general who sent the Hun army running at the Battle of the Chalons. Count Odo of Paris, who we'll meet in a later chapter, commanded a mere 200 men but successfully fought off a vastly larger Viking attack on Paris in 885. Even the Mongols, the most fearsome and successful of all history's invaders, were once turned back by a now nearly forgotten Korean general named Pak So.

But mostly it's the aggressors, not the defenders, whose names are remembered. In the pages that follow, we'll look at the lives and times of some of history's most successful invading hordes, as well as those who defended against them, and in so doing learn how to construct defensive strategies to protect our own castles.

OUR CAST OF CHARACTERS

Alexander of Macedon left his homeland in northern Greece with a large and well-disciplined army in 332 BCE. He marched east out of Thrace and raced through the Levant, Asia Minor, into Persia and beyond to India. Not once was Alexander's army defeated in battle.

The Mongol leader Genghis Khan was no less successful. His warriors captured a portion of the world even greater than that amassed by Alexander. From Eastern Europe to China, no person in the history of the planet commanded as great an area in terms of square miles or percentage of the world's population.

Following in the footsteps of the Mongol khans was Timur or, as he is better known, Tamerlane. He is remembered today as a brutal and predacious conqueror, unrivaled by any other in his cruelty to those he fought and defeated. The monuments he built to commemorate his successes were, like those the ancient pharaohs, giant pyramids. But unlike the brick pyramids of Cheops and Djoser, Tamerlane's pyramids were constructed of human skulls.

Western Europe in the Middle Ages couldn't catch a break. Apart from the periodic waves of aggression from its east, perpetrated by one or another group of Central Asian horsemen, it was equally threatened by seafaring raiders from the north. The Danes, or Vikings as they were usually called during their lifetimes, launched attack after attack on cities, towns, and hamlets in Ireland, England, France, Germany, and Spain. The Vikings were excellent sailors, skilled in hand-to-hand combat, and clever enough to choose easily accessible and poorly defended waterside targets. The effectiveness of these tactics earned the Vikings a formidable reputation as raiders and pirates.

In the 12th and 13th centuries came irregular waves of Christian invaders moving eastward through Eastern Europe toward Jerusalem. These were the Crusaders. While in some cases the Crusaders were chivalrous and valiant, in many other cases they were a sorry lot indeed. The bad ones (and there were many) possessed a rare type of cruelty bred from generations of crushing poverty, ignorance, and superstition. Far too many of the ill-armed, unsophisticated, and un-chivalrous Crusader warriors partially compensated for their lack of sophistication with an intense enthusiasm for rapine and plunder.

Of all the great invaders, least is known about the first great barbarian intruder into Europe. Attila was the king of the Huns, a now-little-known group of nomadic Eurasian invaders who came out of the steppes to threaten, and in some cases crumble, the ruling powers of Western Europe. Like the Macedonians of Alexander before and Mongols of Genghis after them, the Huns rarely lost a battle. But unlike Alexander and Genghis, once gone, Attila left little to remind the world of his presence.

We'll meet each of these worrisome groups, plus a few more, in hopes that understanding their history, armaments, and motivations will help prepare us for our own future battles.

1.3 Most modern homes, like this typical suburban dwelling, are totally unprepared for defense against attack from an invader.

GENERAL SAFETY GUIDELINES

Before you get started, there are some important things you need to know.

1. The projects described here run the gamut from simple to complex. Note that the purpose of many of the projects is to build a device that hurls, shoots, or throws something. Projects should always be supervised by adults.
2. Read the entire project description carefully before beginning the construction process. Make sure you understand what the project is about and what you are trying to accomplish. If something is unclear, reread the directions until you fully comprehend it.
3. Some of the projects call for the use of hand or power tools. Use tools according to manufacturer recommendations.
4. Use care when operating, installing, aiming, and firing these projects. Seek professional guidance when necessary. Use tools, materials, and other gear in strict accordance with the manufacturer's instructions.
5. Use extreme care when using, building, or moving heavy objects.
6. Wear protective eyewear, gloves, and other safety equipment when appropriate.
7. Individual projects may also have their own specific safety instructions.

The instructions and information are provided without any promise or guarantee of safety. Each project has been tested in a variety of conditions. But variations, mistakes, and unforeseen circumstances can and do occur; therefore, all projects and experiments are performed at your own risk! If you don't agree with this, then put this book down—it is not for you.

2

GENGHIS KHAN AND THE MONGOLS

When you're dealing with some of history's most fascinating characters, there's no reason to ease into things like a swimmer wading into an April sea. No, let's plunge right in and explore the lives and times of the most ferocious and devastating invaders in world history: the Mongol hordes that swept into China, Persia, Russia, and Eastern Europe in the 13th century.

How bad were they? Really, really bad. In his bestselling book *The Better Angels of Our Nature*, Steven Pinker says that the Mongol invasions of the early 13th century were perhaps the most vicious and murderous acts of human aggression ever perpetrated by one group of people upon another. Adjusted for world population differences, Mongol attacks killed *three times* more people than the First World War, the Second World War, the Napoleonic Wars, the Russian Civil War, and the American Civil War *combined*.

The Mongols smashed into and destroyed every army they fought. They fought and crushed without exception: the Russians, the Turks, the Indians, the Arabs, the Hungarians, the Chinese, the Koreans, and the Persians, to name just a few of their better-known opponents. City walls, no matter how thick, could not stand up to them, and no fortress, no matter how solid, could provide protection for more than a few fleeting weeks.

There was great motivation on the part of the besieged to resist. When Mongols conquered a city, they frequently assembled the inhabitants in a wide, open area outside the town. Then, each Mongol soldier took his battle ax and killed 10 to 50 unarmed villagers. Apparently, this was higher than the average uneducated, illiterate Mongol warrior could count, so to keep track of the killings, the Mongol would cut off the right ear of each victim, place it in a sack, and take it back to his commander to be counted. At the Battle of Liegnitz in what is now southern Poland, the victorious Mongols filled nine large sacks with ears.

Such was the situation facing the inhabitants of Bukhara, then one of the largest and most prosperous cities in the empire of the shah of Khwarezm. In the year 1220, the city, today part of the republic of Uzbekistan, at first seemed invulnerable even to Mongol invasion. The city had a large garrison of well-trained, well-armed troops. Surrounding the city was a foreboding barrier wall, and inside was a fortress known as the Ark, a redoubt of exceedingly thick stone walls that stretched around a good portion of the inner city.

Near the Ark was the towering Kalyan Minaret; at 150 feet high, it was the highest man-made structure in all of central Asia and a perfect vantage point for the Khwarezmians to keep tabs on Mongol troop movements outside the city. In addition, after the Mongols threatened nearby cities and towns, the city's ruler had brought more than 20,000 additional troops into the city to reinforce it from Mongol attack. With all of these military advantages, Bukhara was one of the most heavily defended cities in the medieval world.

Outside the city walls, in full view of the wary observers high atop Kalyan, the leader of the Mongol horde, Genghis Khan, sat astride his horse and surveyed the situation. Genghis was the great *khan*, the

2.1 The Ark of Bukhara　　　　　　　　　　　　　　*Stomac*

Mongolian word for commander or ruler. Genghis, whose name at birth was Temujin, was the founder of the Mongol Empire and the commander in chief of its vast army of skilled horsemen.

Temujin was born into a humble family, and he lived his early years in very difficult conditions. But in an amazingly short time, Temujin became Genghis and proved himself a leader of exceptional ability. He was responsible for changing what was once a large number of autonomous and frequently warring nomadic tribes that lived throughout central Asia into a tightly run, well-coordinated empire with himself as supreme leader.

He and his descendants went on to spread Mongol hegemony throughout a great portion of Asia and parts of Eastern Europe by conquering the areas covered by modern-day China, Korea, Iran, Syria, Iraq, the Central Asian countries, and portions of Eastern Europe and Russia. In terms of land area, his was the largest empire the world had ever seen.

Inside the walls of Bukhara, the people were well aware of Genghis and his ways of making war. Since reports of Genghis's awful deeds preceded him, even the presence of the fortress walls and the wall-to-wall cordon of troops inside them gave the city's people little confidence that they would remain safe for long. They had heard from survivors of other attacks in other cities what would happen should the Mongol army enter and take the city. So, they sent a delegation to speak to Genghis and see if a peace agreement could be negotiated.

A deal was made, but it didn't turn out well for the Bukharans.

The next day at the main mosque, in the shadow of the great walls and minaret, the leading citizens of Burkhara were ordered to assemble. Once there, a man—tall, long-bearded, red-haired, green-eyed, and dressed in the armor and clothing of a Mongol chief—climbed up the stairs into the minbar or pulpit of the mosque.

"O people," cried Genghis Khan himself to the assembled citizens. "Know that you have committed great sins and that the great ones among you have committed these sins. I am the punishment of

2.2 Kalyan Minaret   *Anatoly Terentiev*

God. If you had not committed great sins, God would not have sent a punishment like me upon you!"

True to his word, he systematically destroyed the town, with the exception of the Kalyan Minaret. Once they finished plundering, the Mongols forced the townspeople to assist them in attacking other towns. The civilians were ordered to set up the Mongol siege engines, and when the battering rams and catapults were ready, they were made to use them against their own people. At the peak of the fighting, the now enslaved Bukharans were forced to serve as human shields, protecting their Mongol conquerors with their bodies. They were killed by the thousands. The conscripted workers were unable to run away because doing so meant instant death from a Mongol arrow to the back.

The story was the same in nearly every Khwarazmian, Chinese, or Russian city the Mongols attacked. At the great trading city of Samarkand, Genghis's next target after Bukhara, the Mongols slaughtered 50,000 people and took another 100,000 as slaves. After leaving Samarkand, the invaders burned crops, leveled cities, and killed every living human being they encountered. Throughout the time of the Mongol invasions, which included most of the 13th century, in every city the Mongols besieged—Otrar, Baghdad, Kiev, and Caizhou to name just a few—thousands upon thousands of innocent civilians were slaughtered and cities razed. All told, more than 40,000,000 people died in the invasions led by Genghis Khan, his sons, and his grandsons.

2.3 Mongols on Horses   *Staatsbibliothek Berlin*

PAK SO

As powerful as the Mongol hordes were, there are some examples of well-prepared people successfully defending their castle. Although few in number, they do give hope that with careful preparation, even the worst threat imaginable can be toughed out.

When Genghis died in 1227, there was considerable infighting among the Mongol chiefs as to who would assume leadership of their vast empire. The Mongols, especially those in positions of power, were nothing if not ambitious. However, the method of choosing a new leader had been fairly well established by Genghis before he died, and eventually, his son Ogedei was selected as the new great Khan.

Almost immediately after gaining power, Ogedei set to work on several tasks that his father had left unfinished. His top priority was to subjugate all of eastern Asia, from Mongolia to Japan. In 1230, Ogedei rode east from his new capital city of Karakorum with a vast host of horsemen. The first item on his to-do list, perhaps a bit surprisingly given the incredible wealth of the parts of China next door, was to take Korea. But the Mongols knew what the powerful Chinese families who controlled northern China perhaps did not: Korea was a soft spot in the well-protected northern border. Therefore, taking Korea was a key goal for Ogedei in his plan to subjugate China.

Ogedei split his forces into two groups, giving his brother Tolui command of an army that went south to take on the Chinese at Kaifeng, which is present-day Beijing. In the north, he appointed his trusted general Sartai to conquer Korea. Sartai was an experienced and battle-tested leader. Sartai's horsemen swept down from the north across the Yalu River and invaded the northern provinces of Korea.

On the Korean side was Pak So, the military commander of the Korean army. Despite the terrifying knowledge of what the Mongols did to resisters, instead of meekly submitting, Pak So mobilized his army and prepared for battle.

In the autumn of 1231 the Korean army was resting after a long day's march when some observers perched on a high hill shouted: "The Mongols are arriving!" It was an army estimated at 8,000 soldiers. Pak So's forces sprang into action and by quickly closing ranks were able to repulse the surprise attack. Even General Sartai's fast and mobile cavalry, the pride of the Mongol forces, ran into such stiff resistance from the Korean infantry that they too retreated.

After the battle, Pak So's forces assembled in the walled fortress of Kuju in the city of present-day Kusong, North Korea. Knowing that the Mongols, with their superior numbers and battle-hardened soldiers, would soon return, the defenders dug in for a fight. It was here at Kuju, a place with large walls and good defensive cover, that Pak So and his men would make their stand.

The regrouped Mongol army wasn't long in coming. They attacked from the west, south, and north. According to the Korean chroniclers of the battle, the fighting was unbelievably intense.

The Mongols besieged the city, attacking day and night. They loaded carts with grass and wood, lit them on fire, and turned them over by the gates as they advanced to attack. In response, the Koreans, it was alleged, used catapults to hurl projectiles at the attackers outside the wall to force them back.

Next, the Mongols constructed great wooden siege towers wrapped with cowhide. They pushed the towers to the gates. Once there, miners hidden inside began to undermine the base of the city walls by excavating a tunnel. But the Koreans countered by boring a hole into the tunnel and pouring molten iron on the miners.

The Mongols had never run into a group quite so well prepared. After a short discussion, they decided to attack the south wall of the city with 15 large catapults. These also met with little success as the Koreans built their own catapults on the ramparts of the city walls and hurled stones back at the Mongols, driving them off. Growing ever more desperate and angry, the Mongols soaked hunks of wood with human fat, reputed to be rendered from the bodies of their victims. Once afire, such projectiles burned unquenchably and could not be extinguished with water.

For a month the Mongols besieged Kuju, attacking it hundreds of times. But regardless of when or how they attacked, Pak So and his men were able to adapt and defend the city. Bringing up reinforcements, Sartai tried again, this time with a whopping 30 catapults. The constant bombardment by these machines opened large holes in the city walls. Valiantly, the Koreans leapt into the voids and repaired the holes by placing iron bands across them.

The Mongols then built scaling ladders and directly assaulted the city. The city defenders countered with huge wheeled knives, or what contemporary observers called "large implements for slashing." All of the scaling ladders were destroyed.

During the siege, a Mongol commander said to be 70 years of age (a very old man indeed for a Mongol) toured the city walls to inspect his troops and equipment. He turned to his men and said: "I have followed this army since I was a youth and so I am accustomed to seeing the cities of the earth attacked and fought over. But I have never seen a city undergo attack like this which did not, in the end, submit."

The other Mongol commanders agreed.

"This city has withstood many with few," said Sartai. "Heaven protects it, not the strength of men."

With this, the Mongols lifted the siege and left.

Hudun Pao: The Crouching Tiger Catapult

2.4 Crouching Tiger Side View

While there's nothing better for knocking down a wall than a catapult, Pak So and his soldiers showed how effectively catapults could be used for *defending* a castle as well.

The Hudun Pao, or Crouching Tiger, is one of the simplest catapults that you can make, but it's powerful and accurate. One of its key characteristics is that the throwing arm rotates on a vertical plane until it smacks hard against the machine frame at the end of the swing. This arrangement means that your war engine packs a wallop. It also requires that you build the frame very solidly. This model can be easily scaled up or down to suit nearly any defensive or offensive job, from a science fair project to turning back attacking Mongol hordes.

A bit ironically, the Crouching Tiger was originally a Mongol design. The rotating throwing arm was powered by a falling weight, and it was so effective that it was adopted by many Chinese and Korean military engineers. Over time, they modified it to make it shoot farther and better, just as we are doing. Our modification uses elastic cords instead of a falling bucket of rocks for power.

The plans that follow detail the construction of a relatively small machine, capable of flinging 3-inch water balloons and other small projectiles across and even beyond the typical backyard.

In case of an actual Mongol attack, you could scale up this project by using larger hunks of wood and substituting large steel springs for the elastic cords.

MATERIALS

☐ Safety glasses
☐ (2) 2-inch × 8-inch boards, 36 inches long: Sides
☐ (1) ¾-inch-thick plywood board, 10 inches × 24 inches: Stop
☐ Wood glue or staples
☐ (1) 6-inch × 6-inch foam pad
☐ (1) 1½-inch-diameter PVC pipe, 43 inches long: Lever Arm
☐ (2) Steel eye bolts ¼-inch diameter, 2-inch-long shaft (Eye bolts are threaded bolts with an eye at the end for attaching a hook.)
☐ (4) Nuts and washers
☐ (2) 1½-inch-diameter PVC pipes, 6 inches long: Pivot Arms
☐ (1) 1½-inch-diameter PVC tee fitting
☐ PVC cement and primer

☐ (6) 2½-inch-long deck screws

☐ (2) 2-inch × 8-inch boards, 12 inches long: Ends

☐ (3) 1½-inch-diameter PVC end cap fittings

☐ (1) #8 machine screw, 1½ inches long, nut, washer

☐ (3) Steel screw eyes, ¼-inch-diameter, approximately 2-inch-long shaft (Screw eyes are similar to eye bolts except the ends are woodscrews instead of threaded bolts.)

☐ (4) Elastic (bungee) cords with hooks, nominally 20 inches long

☐ (1) Trigger [see Step 11]

☐ Small water balloons or other small projectiles

☐ Threadlocker (e.g., Loctite)

TOOLS

☐ Table saw or wood saw and 3-inch-wide wood chisel

☐ 2-inch hole saw

☐ Electric drill with ¾-inch spade bit, #18 twist drill, and Phillips head screwdriver bit

DIRECTIONS

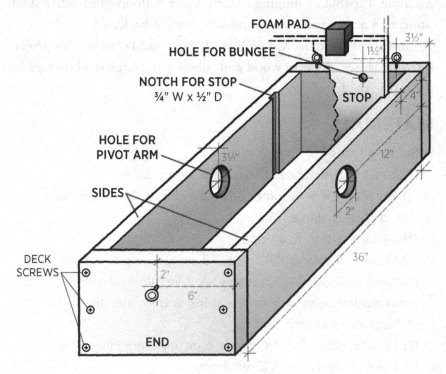

2.5 Frame Assembly

22

1. Put on safety glasses. Use the table saw or wood saw and chisel to cut a ¾-inch wide by ½-inch deep straight groove into the face of each 36-inch-long board, 4 inches from one end. These boards will be the Sides of the frame.

2. Use the hole saw to cut a 2-inch-diameter hole in the face of both Side boards, about 12 inches from the end and 8 inches from the grooves, as shown in **diagram 2.5**. Drill two ¾-inch holes in the plywood board that will be the Stop, as shown in the diagram. (You will insert the bungee cords through these holes in a later step, so you may want to make sure the ends of the cords will pass through the hole. If not, use a larger drill bit.) The holes should be 2 inches from the top and 1.5 inches from the sides. Glue or staple the foam pad to the center of the top of the Stop.

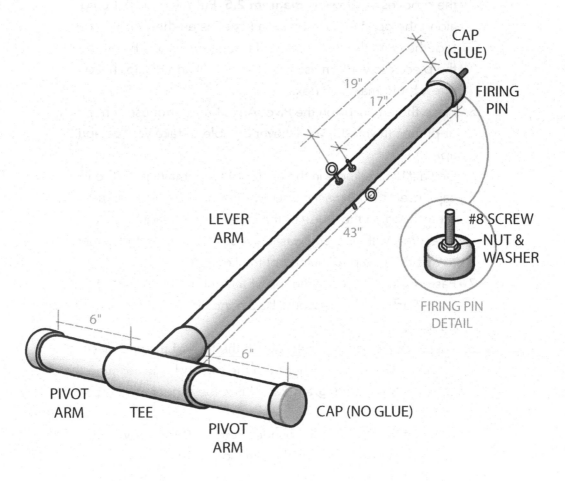

2.6 Throwing Arm Assembly

3. Drill holes for the eye bolts in the 43-inch PVC Lever Arm at 17 and 19 inches from the end, as shown in **diagram 2.6**. Insert the eye bolts and fasten securely with washers and nuts. Be sure to position the bolt's loop upward on the bolt closest to the firing pin end and position the bolt's loop downward on the bolt closest to the Pivot Arms. Build the throwing arm assembly by connecting the Lever Arm, the Pivot Arms, and the tee fitting using the PVC primer and cement as shown in **diagram 2.6**. Do not attach end caps at this time. Read and follow the directions on the PVC primer and cement containers, including information on using it safely.

4. Line up the sides so the grooves face inboard and insert the Pivot Arms in the 2-inch holes. Insert the plywood board into the grooves as shown in **diagram 2.5**. Put the 12-inch boards along the open ends to create a box. Fasten the wood pieces together with the deck screws. If necessary, have a helper hold the pieces so you can use the electric drill and Phillips head screw bit to make the box.

5. Place the end caps on the Pivot Arms but do not use cement or primer. (If you do, you'll never be able to take your catapult apart.)

6. Drill a #18 sized hole in the center of the remaining PVC end cap. Insert the #8 machine screw through the hole with the screw head on the inside of the cap. Using threadlocker, fasten it securely with nuts and washers. Use PVC primer and cement to attach the cap to the end of the Lever Arm.

7. Fasten 2 screw eyes to the top of the end piece closest to the Stop. Each screw eye should be 3½ inches from the corner of the box.

8. Fasten the remaining screw eye to the face of the end piece opposite the Stop, 2 inches from the top and 6 inches from the side, as shown in **diagram 2.5**. This is the trigger screw eye for step 10.

9. Connect the hooks on the bungee cord ends to the eye bolt on the Lever Arm, through the corresponding holes on the stop, and hook the remaining end through the screw eyes on the end.

10. All that remains is the trigger. You can build your own release or use an archer's arrow release from an archery supply store (Internet search term: "archery arrow release"), a pelican hook from a sailing supply store, or a horse trainer's panic snap from a tack shop (search term: "horse panic snap"). Attach one end of the selected trigger mechanism to the trigger screw eye.

2.7 Trigger Attachment

OPERATING THE CROUCHING TIGER

1. Carefully pull the throwing arm back. If the tension on the arm is too great or too little, you can add or remove bungee cords. Don't go overboard, however, as too much stress could break the machine! Latch the lever arm to the archery release, pelican hook, or panic snap. (You may need to use a short loop of rope or a carabiner to attach the trigger release to the trigger screw eye depending on the size and shape of the trigger selected.)

2. Tie a small water balloon or other projectile to a string loop and place the loop over the firing pin.

3. Release the arm and watch your projectile fly!

Safety Notes

1. Depending on the strength and number of bungee cords used, the lever arm can strike the stop with great force. Don't skimp on padding! Make sure there is sufficient padding on the stop to prevent the arm from breaking when it hits. Also, don't use too many bungee cords or stretch them excessively.

2. Keep hands, face, and other body parts well away from the plane of rotation of the throwing arm, especially when the machine is cocked and ready to fire. Be very careful of that rotating arm!

3. Do not stand in front or behind the plane of rotation of the arm. Use this machine outdoors, in areas where the projectile will do no harm.

4. As always for projects of this nature, wear safety glasses and use common sense.

─ ─ ─ ─ ─ ─ ─ ─ ─ ─ ─ ─ ─ ─

THE BEST WAY TO TURN BACK MONGOL HORDES

In the mid-13th century, a Franciscan friar named Giovanni da Pian del Carpine, or John de Plano Carpini as he was called in English, traveled to Mongolia to visit the court of the great khans. Granted admittance as an ambassador of the pope, he took every advantage of the situation, making copious notes and drawings of what he saw. Upon his return to the West, he wrote Pope Honorius III a long letter saying that based on what he saw and heard, Europe should soon expect a Mongol attack. He went on to advise his superiors that there might be one particularly effective weapon in turning back the Mongols: the crossbow.

Whoever wishes to fight against the Mongols ought to have crossbows, of which they are much afraid, and a good supply of arrows. The heads of the arrows for the crossbows ought to be tempered after the Tartar fashion, in salt water when they are hot. This makes them hard enough to pierce the Mongol's armor.

2.8 Crossbowman   St. Nicholas Illustrated
Magazine, *New York, 1887*

Although the crossbow was first devised and used by the Roman army, it disappeared from the armories of most Western countries for nearly 700 years. But once it was rediscovered, it was a game changer and was quickly adopted by armies because it was powerful, accurate, and didn't require years of practice to be effective, as did the longbow. It was made out of a composite spring (the bow) that consisted of multiple layers of animal ligaments and wood fibers, all laminated together with glue made from river sturgeon. Starting at about the time of the Battle of Hastings in 1066, the crossbow reappeared on European battlefields.

According to Sir Ralph Payne-Gallwey, an early 20th century British historian and one of the most well-known scholars on the subject of crossbows and catapults, by the time of the Mongol Invasions, the crossbow was considered the most advanced handheld infantry weapon available.

According to Sir Ralph, learning to use a crossbow was quick and easy—too easy in the minds of some medieval government and religious officials. With a few days of practice, any soldier could learn to load, aim, and fire a crossbow. By contrast, archers using the English longbow, another formidable weapon, required years of practice and very strong arm muscles.

The ease in which one could become proficient with a crossbow led to the first enactment of "gun control" laws. In 1139 the Catholic pope, Innocent III, issued a papal bull that outlawed the use of crossbows. They were, he said, "hateful to God and unfit for Christians." If weapons like this got into the hands of insurgents or heretics, well, that could shake the foundations of government and religion. Fearful of the destabilizing potential of the crossbow, the Christian countries of Europe obeyed Innocent's proclamation and destroyed their crossbows.

This was the time of the Crusades, however, and soon the English and Frankish armies found themselves in pitched battles against Turkish troops who had no such proscription against the use of crossbows. The fusillade of crossbow bolts shot by Turkish defenders made the Crusaders reconsider using the weapon. Eventually, Richard I of England, better known as Richard the Lionhearted, reintroduced crossbow use among the rank-and-file Crusader troops. Ironically, Richard died of gangrene in 1199 from a wound inflicted by a crossbow bolt.

An infantry man proficient in crossbow use was an extraordinarily valuable asset and often commanded higher rates of pay than other foot soldiers. In particular, Genoese crossbowmen were well regarded and were widely hired as mercenaries throughout medieval Europe. Their Italian-made crossbows, each weighing a hefty 18 pounds, could shoot a quarter-pound arrow that could penetrate armor at a range of 100 yards. The bows were equipped with a lever, called a goat's foot, that allowed a bowman to apply leverage so the heavy bowstring could be pulled back and cocked.

Such weapons were devastating devices indeed. As Carpini surmised, even the Mongol horsemen, who were nearly fearless in battle, thought twice or even three times when attacking crossbow-protected fortresses. For instance, during the Mongol invasion of Russia in the year 1240, the Mongols suffered a rare defeat when they were turned away from the city of Kholm by the withering fire from Russian crossbowmen.

Carpini's Crossbow

Designing and constructing a crossbow of military or sporting quality requires great skill in addition to fairly sophisticated metal and wood-working tools. However, with minimum tools and the directions that follow, you can make a model crossbow suitable for defending your castle against "model" (by this I mean imaginary) Huns or Mongols. You're probably thinking that the model will not provide much in terms of actual defensive capabilities. And you're right; it won't. But the concepts used in building it could be expanded upon, and if you are clever, a crossbow with usable power and accuracy could be fabricated, should a real threat emerge.

The model crossbow is an excellent team project, and you'll have a great deal of fun in both making it and using it. Remember, however, that you're building a weapon that shoots a projectile; therefore, using it safely is the most important part of the project. Don't aim at things you don't want to shoot, don't overstress the parts by pulling the bow back too much, and wear protective gear (for example, safety glasses) as appropriate.

MATERIALS

Note on materials: You have quite a bit of latitude in building this project. You can make the stock, bow, and trigger a bit longer or shorter and it will still work OK. The dimensions in the drawings are the ones that worked best for me and provided good results. However, feel free to experiment with the dimensions and perhaps get even better performance!

☐ (1) 1¾-inch × 1¾-inch square dowel, 36 inches long: Barrel (available at most large home stores or lumberyards, or search online for "1¾ square dowel")

☐ (1) 1¾-inch × 1¾-inch square dowel, 12 inches long: Gunstock

☐ Wood glue

☐ (2) Wood strips, ⅜-inch × ¼-inch × ½-inch: Spacers (You can either cut these to size from a piece of scrap pine board or buy a precut piece of basswood at a hobby store.)

☐ Aluminum strip, 1½ inches × 9 inches × ⅛ inch: Trigger

☐ (1) 1½-inch hinge, with mounting screws

☐ Wood strip 1½-inches × 2½-inches × ⅛-inch: Trigger Pad

☐ (1) Loose coil spring ⁹⁄₁₆-inch diameter, about ½ inch long (You probably won't find this exact size at the hardware store, but you can buy one a bit bigger and cut it down to size.)

☐ Axle grease (optional)

☐ (3) #8 short machine screws and nuts

☐ (1) 4-inch-diameter U-bolt, strap, and nuts: Stirrup

☐ (2) #10 screw hooks: Puller Hooks

☐ (1) Round dowel, 1-inch diameter × 6 inches long: Puller

☐ (1) ⅜-inch × 1¼-inch × 36-inch-long oak board: Bow (You can cut this yourself from a board or dowel, but it's easier to buy a piece of moulding this size at the home store. It's called a mullion. Home Depot's Internet part number is 203116469.)

☐ (2) #8 wood screws, 1 inch long (for attaching Bow)

☐ (1) Piece of stiff, inelastic cord about 48 inches long: Bowstring (Four-ply waxed linen cord works very well and most fabric stores sell it. Hardware stores sell #18 mason twine, which works acceptably.)

☐ Small projectiles to fire

☐ (1) ⅜-inch-diameter bolt, 2 inches long: Firing Bolt

TOOLS

- ☐ Safety glasses
- ☐ Hand saw or table saw
- ☐ Sandpaper or file
- ☐ Router with a ¾-inch straight router bit and a ⅜-inch straight router bit (Alternatively, you could use a saw and chisel, but it's more work.)
- ☐ Electric drill with ³⁄₃₂-inch, ¾-inch wood bit
- ☐ Screwdrivers, pliers, vise

DIRECTIONS

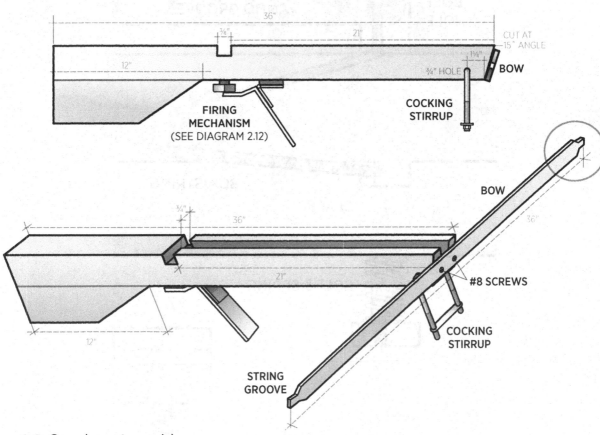

2.9 Crossbow Assembly

1. Put on safety glasses. Use the saw to cut one end of the 36-inch-long square dowel, which will be the Barrel, at a 15-degree angle, as shown in the diagram. Cut one end of the 12-inch square dowel, the Gunstock, at an angle as shown. Glue the pieces together with both flat ends flush and the angles toward the same direction, and let dry. Round the edges of the Gunstock with sandpaper or a file.

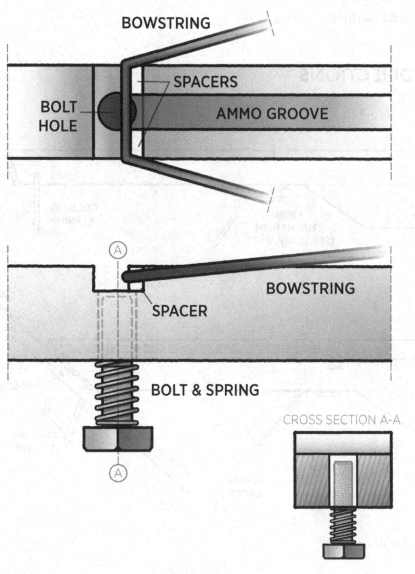

2.10 Firing Mechanism Assembly

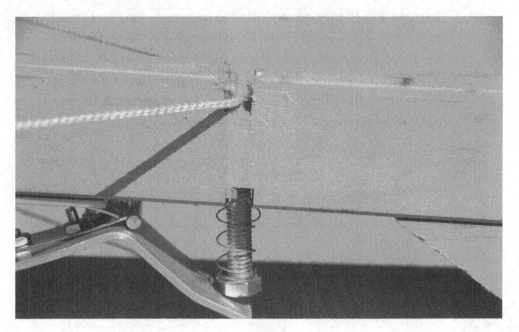

2.11 Firing Mechanism

2. To make the ammo groove in the Barrel, use a router or a saw and chisel to cut a ¾-inch groove, ¼-inch deep, in the center of the top face of the Barrel. This groove holds and guides the ammunition. See **diagrams 2.10** and **2.11**.

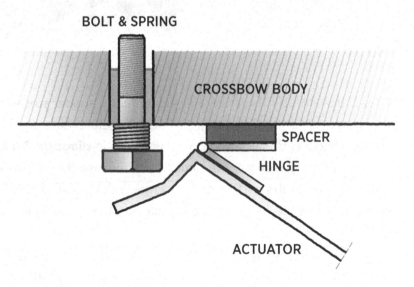

2.12 Firing Mechanism Detail

3. To make the Trigger groove, use the router or a saw and chisel to cut a ⅜-inch wide by ⅜-inch deep groove in the top face of the Barrel perpendicular to the ammo groove, at a point about 21 inches from the angled end. Drill a ½-inch-diameter hole in the exact center of the ⅜-inch groove. See **diagram 2.12** for help in determining how the hole and grooves align.

2.13 Spacer Strips

4. Glue the ⅜-inch by ¼-inch by ½-inch centering spacer strips to the front edge of the Trigger groove, one on either side of the centered hole in the ammo groove, as shown in **diagram 2.13**. The centering spacer strips are important because they position the bowstring in the middle of the trigger groove so the bolt pushes against the center of the bowstring when the Trigger is pulled.
5. To make the Trigger, use a bench vise or the side of a sturdy table to bend the longer aluminum strip as shown in **diagram 2.12**. You will need to experiment a bit with the two angles

on this piece of aluminum to obtain the action required to move the firing bolt upward when the Trigger is pressed. Luckily, aluminum bends easily and is quite forgiving, so you can make several adjustments if necessary. Take your time and experiment until the firing bolt smoothly and dependably lifts and fires the bowstring.

With the woodscrews that came with the hinge, attach one end of the hinge to the Barrel. Depending on the hinge you purchase, you may need to insert a thin wood spacer, made from a piece of scrap, between the hinge and the Barrel, to mount the hinge.

Place the spring around the ⅜-inch bolt and insert the bolt into the ½-inch hole in the Gunstock with the bolt head on the bottom. Make sure the bolt moves up and down easily in the hole. File or sand the interior of the hole and use axle grease if necessary.

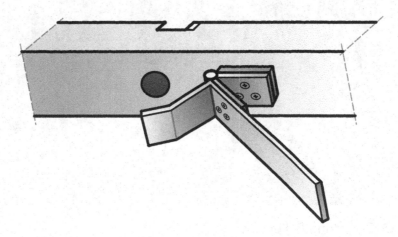

2.14 Trigger Attachment

Drill three holes in the aluminum using the hole pattern on the hinge as a pattern. Then, attach the Trigger to the hinge using three #8 short machine screws and nuts as shown in **diagram 2.14**. When you pull back on the long end of the Trigger, the short end should push the bolt up, which in turn will push the bowstring out of the Trigger groove and fire the projectile.

6. Make the cocking Stirrup by drilling a hole, ¾-inch or larger, depending on the diameter of the U-bolt you are using, in the Barrel perpendicular to the ammo groove, at a point about 1¼ inches from the angled end of the Barrel. Insert the U-bolt and secure with the strap and nuts. Make the Puller by attaching the screw hooks to the 1-inch-diameter dowel 2 inches from either end. Align the hooks so they face the same way. See **diagram 2.15**.

2.15 Bowstring Puller

7. Use the router or saw to cut a string groove in both ends of the ⅜-inch by 1¼-inch by 36-inch-long oak mullion Bow as shown in the circled detail in **diagram 2.9**. Attach the bow to the angled end of the barrel with two #8 screws, 1 inch long. Position the screws so they avoid hitting the U-bolt. Tie two loops in the ends of the bowstring so that it is about 45 inches long. (You can adjust the length of the bowstring later, based on your results.)

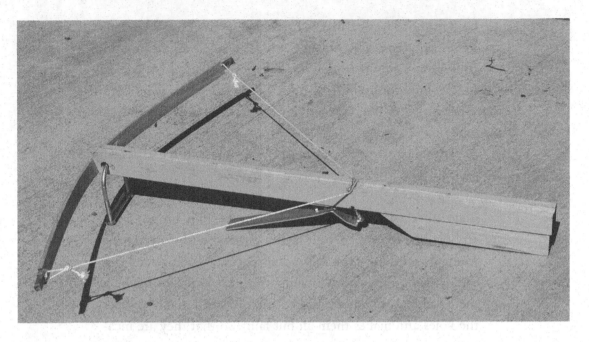

2.16 Completed Crossbow

USING YOUR CROSSBOW

Holding the crossbow so the front end is facing downward, place your foot in the stirrup to steady it. Then, using the puller, pull the bowstring back and down into the trigger groove. Place your ammunition (typically a small rock or arrow, but you can be creative with this) in the groove just in front of the trigger groove. Aim and then pull back on the trigger. Pulling back on the trigger will push the bolt up, which in turn will free the bowstring from the groove. The bowstring will push on whatever ammunition is placed in the ammo groove and propel it forward.

Inspect the crossbow parts for wear and replace as needed. Remember: use common sense when operating your model crossbow and do not aim at things you do not wish to shoot. (In case of actual Mongol attack, you may want to consider using a larger and more powerful method of defense!)

At the Battle of Liegnitz in 1241, the Mongols spread a gargantuan smoke screen across the field of battle. It was produced by burning now unknown items in a strange and demonic-looking smoke projector. Using clouds of smoke that both acted as an irritant and hid their movements, the Mongols' Smoke Monster was a device that could turn the tide of battle.

According to medieval chronicler Jan Dlugosz, this early chemical weapon was a

> huge lance with a giant X painted on it. It is topped with an horrible, ugly head with a chin covered with hair. As the Mongols withdraw some hundred paces, the bearer of this standard begins violently shaking the great head, from which there suddenly bursts a cloud with a foul smell that envelopes the Poles and makes them all but faint, so that they are incapable of fighting. . . . Seeing that the all but victorious Poles are daunted by the cloud and its foul smell, the Mongols raise a great shout and return to the fray, scattering the Polish ranks that hitherto have held firm, and a huge slaughter ensues.

This is one of the first recorded uses of chemical warfare in Europe, and it was a turning point for the Mongols in their war. The monster threw Duke Henry's army of eastern European knights into confusion as the "evil-smelling vapors and smoke" hid the activity of the Mongols from them. The Mongols charged through their opponent's lines and made quick work of them. Contemporary reports say that Henry's army took more than 30,000 casualties, many of whom were butchered after death; their ears were removed and sent back east in giant sacks to the Mongol's main camp as a memento of the great victory. Duke Henry himself fared even worse. After he was killed in battle, his head was removed from his body, placed upon a spear, and carried around the walls of the city.

2.17 The Battle of Liegnitz

- - - - - - - - - - - -

Mongol Smoke Monster

What's sauce for the goose, goes the old saying, is sauce for the gander. So, if smoke weapons worked for the Mongols, it's logical that they would work against them as well. Making life as unpleasant as possible for those attacking you makes a great deal of sense. Presumably, if the smell was bad enough and the smoke thick enough, the attacking horde would decamp and try a different, less resolute target.

The following project shows how to make a simple but effective smoke screen. The amount of smoke produced could be increased as required should a larger and more menacing threat emerge.

MATERIALS AND TOOLS

- ☐ Safety glasses
- ☐ Electronic scale
- ☐ 7 grams potassium nitrate (KNO_3) (Potassium nitrate is inexpensive and easily available on the Internet.)
- ☐ Coffee grinder or mortar and pestle
- ☐ 4 grams powdered sugar
- ☐ .5 gram sodium bicarbonate (baking soda)

- ☐ Plastic mixing container with tight-fitting lid
- ☐ Electric skillet or pot, or heat-proof bowl and electric hot plate
- ☐ Heatproof plastic spatula
- ☐ Waxed paper
- ☐ Candy mold (optional)
- ☐ Spoon
- ☐ Homemade blackmatch or commercially manufactured Visco fuse (Fuse is easy to find; just enter "visco fuse" on an Internet search engine.)
- ☐ Matches or lighter

2.18 Smoke Bomb Materials

1. If you are using granulated or prilled KNO_3, often available from local chemical suppliers or garden stores, grind a small amount into a fine powder with the coffee grinder or mortar and pestle. This coffee grinder should *only* be used for grinding individual chemicals. If you use it for mixing chemicals, don't use it for food. Measure the following using an accurate beam or electronic scale:

   7 grams of potassium nitrate
   4 grams of powdered sugar
   .5 gram of sodium bicarbonate (baking soda)

2. Place the potassium nitrate, powdered sugar, and sodium bicarbonate in a container, tightly close the lid, and shake until well mixed.

3. Find a place outdoors for the following steps. Wearing your safety glasses, place a pot on a carefully controlled electrical (no open gas flame allowed) hot plate heated to 285 degrees F. Add the entire smoke screen mixture and stir continually using the spatula. In one to two minutes, the mixture will soften and then melt into something that looks like runny peanut butter. When that happens, it's ready to be shaped or poured into a mold.

2.19 Smoke Bomb Mixture

4. Turn off the heat (if the mixture overheats, it can ignite). Carefully place the hot, viscous mixture onto the waxed paper. Use a spoon to shape it into a rough triangular mound. Alternatively, you can pour it into a heatproof candy mold.

5. Cut a 2- to 3-inch length of fuse, which should be sufficient to provide approximately four to six seconds of burn time. Insert the fuse into the mixture.

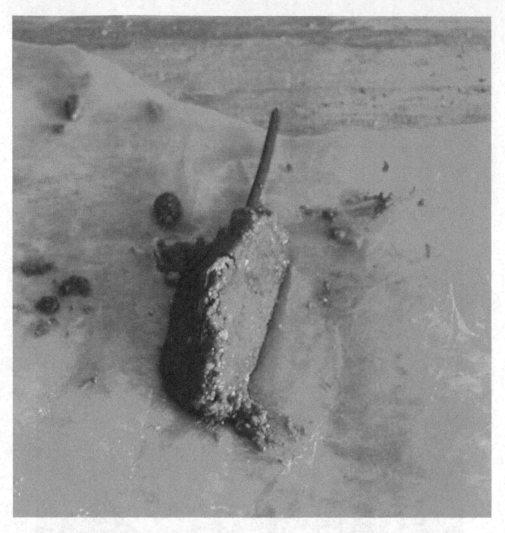

2.20 Finished Smoke Bomb

6. The mixture will harden as it cools. After approximately 15 minutes of cooling, the smoke screen device is ready for use.

7. Place the Smoke Bomb on the ground outside in a safe area, then light the fuse and get away fast. The device you just made deserves a lot of respect! Do not hold it in your hand when lighting it, and do not throw it when it's burning. Do not even think about igniting a Smoke Bomb in a closed container, and don't carry it in your clothing.

3

ATTILA AND THE HUNS

With the possible exception of the Mongols, no group is more closely identified with savage fury and barbarism than the Huns. In particular, their leader, Attila, remains the archetypal king-of-all-brutes, a cold-blooded destroyer of villages and towns, depraved killer of men, and merciless enslaver of women and children. The devastation the Hunnic army spread as it careered through Europe was unprecedented, and Attila's European victims called him the "Scourge of God."

Today, 1,800 years after he laid waste to the largest cities in France, Germany, Italy, and Eastern Europe, Attila is likely the world's best-known least-known person. Everyone has heard of him, but we know very little about him. There are few contemporary descriptions of the Hun people, so their lifestyle and culture are mostly a mystery. We don't really know what they looked like. They had no written language or particularly notable artistic skills, so no records and few artifacts remain for historians and archaeologists to study. What we do know about them comes from accounts of their enemies.

3.1 *The Huns at the Battle of Chalons*

A Popular History of France from the Earliest Times, Volume I

For example, Ammianus Marcellinus, a Roman writer who was a contemporary of Attila, wrote:

> The people called Huns live beyond the sea of Azof, on the border of the Frozen Ocean. They are a race savage beyond all parallel. At the very moment of birth the cheeks of their infant children are scarred by an iron, in order that the hair instead of growing at the proper season on their faces, may be hindered by the scars; accordingly the Huns grow up without beards, and without any beauty. They all have closely knit and strong limbs and plump necks; they are of great size, and low legged, so that you might fancy them two-legged beasts.

No corroborating evidence exists of Ammianus's claims of the Hun appearance, so it's probable that most of his words are mere anti-Hun propaganda, designed to frighten and unify the Romans against the Huns.

Still, it's probably true that the Huns' reputation for ferocity was well earned. Consider, for example, one of Attila's raids in the Balkans during the 440s that occurred at or near the modern-day Serbian city of Nis, then called Naissus. The Hun army laid waste to the city. Historical accounts of the day, all written by Attila's archenemies, say that the city was so devastated that several years later, the river banks were still filled with human bones. Visitors to the ruined city found the stench of rot and decay so great that the place was completely uninhabitable. Whether completely true or not, it is certain that the Huns destroyed Naissus and many other European and Asian cities.

An excellent account of Attila's rise and fall was written at the turn of the 19th century by University of London historian Sir Edward Shepherd Creasy. He writes in his book *The Fifteen Decisive Battles of the World* that starting in the fourth century CE, the power of the Roman Empire was in decline. Roman rule was disintegrating, and German-speaking tribes took control of many of the most prosperous and fertile regions of the Roman Empire. The Visigoths held the north of Spain and France south of the Loire River. Other tribes, including Franks, Alans, and Burgundians, had established themselves in the French provinces, and the Vandals controlled North Africa. Most powerful of all "barbarian" groups were the Visigoths, led by King Theodoric. Rome was contracting under pressure on nearly every one of its borders.

But Rome's problems were about to become even worse. In the areas in and near present-day Mongolia, a huge group of nomadic horsemen began pressing westward. These people, the Huns, had long skirmished with the Chinese, but now they decided to turn their horses around and wheel toward Europe. In 375 CE the Huns crossed the Don River and leapt from the central Asian steppe into Europe.

Once in the west, they fought and conquered every tribe and ethnic group they met. They defeated many powerful barbarian groups including Alans, Ostrogoths, and Lombards. The Roman emperor, closely tracking the progress of this new and powerful threat, became alarmed at their swift progress and brutal tactics. Seeing them heading for Rome, he ordered his legions to check their progress. But the Huns rode roughshod over the legion sent to stop them. In short order, most of Rome's provinces south of the Danube River were lost to the wild Asiatic invaders.

Possessing neither the temperament nor the agricultural skills that would permit them to stay in one place for long, the Huns kept moving west. Town after town, people after people fell to the swift, short-statured horsemen from the Central Steppes. Their leader, Attila, guided his army with a level of precision not seen previously.

When the legends are stripped away it's apparent that Attila was not just one of many of Rome's barbaric conquerors. Sir Edward Shepherd Creasy, who analyzed the works of many ancient chroniclers, concludes that while brutal and coldhearted, Attila possessed finely developed instincts for battle command as well as a thorough knowledge of tactics and strategy. In addition, wrote Creasy, Attila wasn't all sword and spear. While his command skills were unmatched, he also included statesmanship and even well-developed interpersonal skills to gain his political goals. In short, he was able to gain influence through the affection of his friends and the fears of his foes.

"Austerely sober in his private life, severely just on the judgment seat, grave and deliberate in counsel," wrote Creasy,

> he was rapid and remorseless in execution. He gave safety and security to all who were under his dominion, while he waged a warfare of extermination against all who opposed or sought to escape from it. He watched the national passions, the prejudices, the creeds, and the superstitions of the varied nations over which he ruled, and of those which he sought to reduce beneath his sway: all these feelings he had the skill to turn to his own account.

As the Huns approached Central and Western Europe in the mid-fifth century, it's not hard to imagine the terror the inhabitants felt when they heard the news that Attila had founded a new capital on the Danube. The question had become when, not if, he would invade.

In 450 CE, Attila prepared his vast military to take Western Europe by force. The Hun army, already large, was swollen by recruits from the tribes they had subjugated. Roman historians of the day wrote that Attila's army numbered over 500,000 men. This is too large a number to be credible, but it is certain that by the standards of the day, Attila's army was very large indeed. He began his march, sweeping north and west from his new capital, and crossed the Rhine River. He took on and defeated German tribes, who decided to fight him in pitched battles on open fields. It was a bad idea, as the cities of Worms, Mainz, Cologne, and Reims were reduced, sacked, and looted.

FLAVIUS AETIUS AT CHALONS

Attila continued his way through Gaul, now called France, his army devouring and destroying nearly every city in its path. The Huns bypassed Paris, at that time an unimportant hamlet on a river, and marched toward the major city of Orleans. But before they got there, they were met in battle by a combined army of Roman soldiers and Goth warriors. Under the command of General Flavius Aetius, the pivotal fight, known today as the Battle of Chalons, was intense and incredibly bloody. At the end, the Huns lost. Attila and his weakened army figured it was time to head home.

Still, the huge losses at Chalons did little to deter Attila. During the long march back to the Danube and home to the steppe, the remaining Huns enthusiastically continued their fighting and pillaging. During the retreat from Orleans a Christian hermit approached Attila, saying, "You, Attila, are the Scourge of God, brought to Earth for the chastisement of the Christians."

Attila thought it over. Yes, perhaps he really was an agent of this hermit's God, and maybe he was put on Earth to visit vengeance upon the sinful people of Europe. Attila told his followers and his enemies that this would be his new title. Attila, the Scourge of God, became the manner in which he became most widely known.

A few years later, Attila died and was replaced by his son Ellac. In 454 CE the Huns, led by Ellac, were soundly defeated at the Battle of Nedao by an army composed of Goth, Gepid, and Alan fighters, the same peoples that Attila had whipped soundly several years earlier. The Huns never returned to great power and faded from history.

The Palisade Wall

When an earthquake destroyed the city walls of Constantinople in 447, the timing couldn't have been worse. Emperor Theodosius II had just received word that Attila and the Huns were on their way to attack the city. He called together his advisors, and they devised a plan to persuade the city's sports teams to have a race in rebuilding the walls quickly. It worked. In just 60 days the walls were rebuilt, barely in time to turn back the Huns.

Of course, a sturdy defensive wall can be made in even less than 60 days if necessary. A wall made of closely spaced, reinforced timbers, set in the ground and sharpened on the top, is called a palisade wall. The word *palisade* comes from the Latin words *palicea* and *palis*, meaning a stake. The palisade wall was one of the earliest defensive structures ever used. Numerous archaeological excavations have revealed that beginning in the early Bronze Age, people used tools with sharp edges to fashion palisade walls in many places around the world.

Palisade walls were erected for defensive positions from the times of Neolithic hunters through the late 19th century. That's nearly 5,000 years of palisade wall building. Few building structures were used as long in such a relatively unchanged fashion. Walls like these had a lot of advantages: the materials were easily found, the construction techniques required were not hard to learn, they were quick to construct and, once up, provided a modicum of security against infantry and horse-mounted attackers.

3.2 Palisade Wall

MATERIALS

Note: This is general information only. The quantities required depend on the length of the wall. Eight 4- to 6-inch-diameter trees will construct a section about 4 feet long.

- ☐ Logs, 12 to 15 feet long and 4 to 6 inches in diameter; the number of trees needed is calculated by dividing the length of the wall required by the average diameter of the trees: Pickets
- ☐ 2-inch × 4-inch boards, 4 feet long; the logs are nailed to these boards to form 4-foot sections. Determine the number of boards by measuring the total wall length planned and multiplying by 4: Section Boards
- ☐ 4-inch-long nails
- ☐ 2-inch × 4-inch boards, 2 feet long; these boards connect the sections. You'll need the same number of connector boards as Section Boards: Connector Boards

TOOLS

- ☐ Axe or chainsaw
- ☐ Adze or drawknife
- ☐ Hammer
- ☐ Shovel
- ☐ Heavy rope, front-end loader, or gin pole crane

DIRECTIONS

Making a palisade wall is simple, but care must be taken to build the wall so that the spaces between Pickets are minimized. Choose logs for straightness.

1. Prepare the Pickets. Use the axe or chainsaw to remove limbs from trees and cut the logs to uniform lengths. Debark the trees with an adze or drawknife. Using the axe or chainsaw, sharpen one end of each log to a point. Move the Pickets to the place where the palisade will be built.

2. Build wall sections out of Pickets. Lay the Pickets on the ground and arrange them into wall sections 4 feet long. Lay the 4-foot-long 2 × 4 boards across the sections at points approximately ⅓ and ⅔ of the way down the Picket. Affix the boards to the Pickets with nails.

3. Use the shovel to dig a square trench along the area to be enclosed by the palisade. The trench should be 1 foot wide by 3 feet deep.

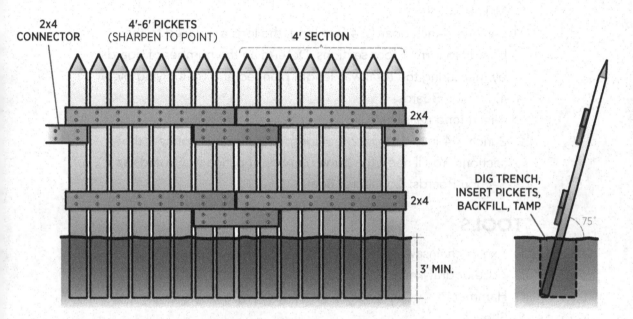

3.3 Palisade Wall Construction

4. Because the wall sections are very heavy, it will take a lot of people to place the wall sections in the trench. Position the sections so the picket points face outward. With 12-foot-long pickets, the palisade will be tilted at 15 to 20 degrees from vertical in a 3-foot hole, making the wall about 8 to 9 feet high.

    Assuming the section is made from pine or a wood of similar density, each section of 12-foot poles will weigh around 800 pounds. That's a lot to lift by hand. To tilt, move, or position the sections, you'll need help. A front-end loader or similar machine will help immensely with this task.

BUILDING A GIN POLE CRANE

But what if you don't have access to power machinery? You can place the poles individually in the trench and attach the Section Boards and Connector Boards after the Pickets are in place. Keep in mind that even a single 12-foot pole weighs about 100 pounds.

Alternatively, you can set up a piece of rigging equipment called a gin pole. A gin pole is a strong, slightly leaning mast or pole. It is stabilized for lifting by placing the gin's bottom in a small hole and attaching four sturdy, well-anchored guy lines to the top to stabilize it. With pulleys at the top and bottom, it is a sort of poor man's crane. But a gin pole can handle a lot of weight: a 6-inch-diameter, 20-foot-long rig can handle more than 2 tons.

MATERIALS

- ☐ (1) 3-inch × 3-inch × 8-foot-long landscape timber, round: Gin Pole
- ☐ (4) Pine blocks 2 inches × 2 inches × 1 inch: Cleats
- ☐ (12) 8d nails
- ☐ (1) Large screw eye
- ☐ 100 feet Manila hemp rope, ⅜-inch diameter
- ☐ Triple pulley block and tackle with at least 60 feet of rope (These are available inexpensively at places such as Harbor Freight, Tractor Supply, and Amazon): Hoist
- ☐ (8) Wooden dowel, 1 to 1½ inches in diameter, 24 inches long: Stakes

TOOLS

- ☐ Shovel
- ☐ Knife
- ☐ Hammer
- ☐ Saw
- ☐ Mallet or sledgehammer

1. Use the shovel to dig a hole 6 to 10 inches deep a short distance from the load to be lifted. The distance from Gin Pole to load should be no more than one-third the length of the pole.

2. Attach Cleats to the Gin Pole with three nails per Cleat, as shown in the diagram. The Cleats will keep the ropes from sliding down the pole. Attach the screw eye to the pole about 1 foot from the lower end.

3.4 Gin Pole Cleats

3. Next, connect the Hoist by lashing the upper block and tackle hook to the top of the Gin Pole just above the lower Cleat. Loop the rope around the hook three times and then use a secure knot such as a clove hitch or timber hitch to tie off. (See chapter 7 for instructions for making these knots.)

3.5 Gin Pole Hoist Attachment

4. Use a knife to cut four guy lines about 20 feet long from the 100-foot-long rope. Place the Gin Pole on its side and tie four clove hitch knots just above the upper Cleat. Tie the knots so that when each guy line extends straight out, the rope is not bent or doubled back, as shown in the photograph.

5. Locate the Stakes 9 feet from the hole at compass points of 45 degrees, 135 degrees, 225 degrees, and 315 degrees from the line extending between the load and the hole. Pound in the Stakes until about 8 inches remain exposed to view, at an angle about 25 degrees from vertical, extending away from the pole.

3.6 Gin Pole Stakes

3.7 Gin Pole in Action

I can tell you that lifting heavy objects, like a palisade fence section, can be a dangerous business if done improperly. Inadequate staking, poorly tied knots, using undersized masts, or old, weak ropes can result in a broken rig and a bad situation. Rigging is an art, so start small and increase your loads slowly as you become more proficient. And *never* be under the load or mast during a lift.

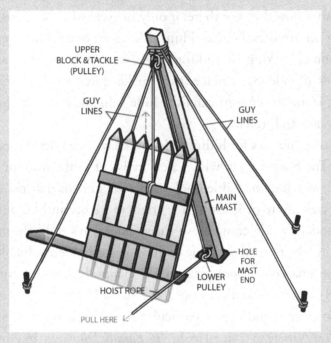

UPPER BLOCK & TACKLE (PULLEY)

GUY LINES

GUY LINES

MAIN MAST

HOLE FOR MAST END

LOWER PULLEY

HOIST ROPE

PULL HERE

3.8 Hoisting the Palisade Wall

5. Connect the wall sections into the full-length palisade wall using the 2-foot Connector Boards to connect each wall section to its neighbor as shown in the diagram.
6. Backfill the trench with soil and tamp down the soil firmly to solidify as much as possible.

Your palisade wall is complete. You can use a chainsaw or rip saw to cut holes for observation as needed.

The Hidden Book Safe

Sometimes it's better to be clever than strong. That's frequently the case when it comes to safeguarding small valuables within your home. Despite your best efforts, it is possible that the barbarians will eventually make it into your castle. Perhaps you weren't home to defend it or you beat a strategic retreat. But when you've got Huns in your home, it doesn't matter how they got there, it only matters what happens next.

Barbarian invaders, be they Huns, Mongols, or anyone else, have no intention of sticking around for long after they've taken what they want. After plundering a place, they make quick work of leaving. Therefore, it makes sense to carefully hide valuable items so they are not discovered and stolen.

Various techniques for hiding valuables have been developed over the years. The book safe, in which a cavity is cut in the interior pages of an otherwise unremarkable book, is so well known that many pillagers are well aware of it. Still, it is tried and true, and because the average modern castle contains so many books, it's unlikely that an invader will take time to examine each book for hidden valuables.

Choose a hardcover book, at least 1½ inches thick with a trim size of 5½ by 8 inches. While it's not necessary, many book safe makers have a wry sense of irony and choose particular books, such as Adam Smith's *The Wealth of Nations* to conceal cash, *The Eustace Diamonds* by Anthony Trollope to hide jewelry, or any Zane Grey novel to stash a handgun.

At first glance, cutting a cavity in the pages of a book seems so easy and straightforward that no directions are required. Armed with a sharp X-ACTO knife and plenty of spare blades, one can cut a reasonable stash area in any book. But it takes a lot of time this way, especially if the cavity is more than half an inch deep. In addition, using a utility knife to cut a few pages at a time invariably yields a rough and uneven cut. If you are patient and diligent it is certainly possible to make a fine book safe this way, but a rushed job will cause crinkled pages and may tip off the thief or plunderer that something is up with this book.

A faster, neater, and all-around better method makes use of a jigsaw, a plywood form, and clamps. The rigid form makes cutting a neat cavity simple and easy.

MATERIALS

☐ Hardcover book, at least 1½ inches thick, 5½ × 8 inches

☐ Marker

☐ (2) pieces ⅜-inch-thick plywood, cut to the same width and height as the book's pages

☐ 1-inch paint brush

☐ White glue

☐ Heavy object, such as a brick or barbell weight

TOOLS

☐ (2) 4-inch C-clamps

☐ Electric drill with drill bit diameter just larger than the size of the jig saw blade

☐ (2) ⁵⁄₁₆-inch bolts, an inch longer than the thickness of the book, with nuts and fender washers

☐ Jig saw

1. Use a marker to mark off the length and height of the cavity desired on one of the pieces of plywood.

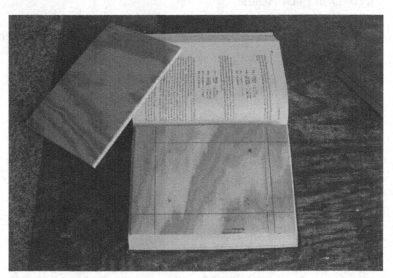

3.9 Marking the cavity size

2. Choose the appropriate number of pages for your cavity. If you select a large number of pages, the cavity will be deeper and you can hide larger items in the book safe. However, paper is

surprisingly difficult to cut through, so some smaller jigsaws will not be able to do the job.

Place the plywood forms on either side of the selected pages. Clamp the paper and plywood forms together securely with the C-clamps. The rigid form makes it easy to cut a clean, uniform cavity.

3. Use the electric drill to drill pilot holes for the jigsaw blade in opposite corners of the cavity layout drawn on the plywood.

3.10 Drill Pilot Holes

4. Drill 2 ⅜-inch holes in the center of the plywood forms. Insert the ⁵⁄₁₆-inch bolts, using fender washers on either side, and tighten the nut, as shown in **diagram 3.11**.

3.11 Fender Washers

5. Use the jigsaw to cut out the paper cavity.

3.12 Cutting Paper Cavity

6. With a brush, spread white glue on the interior of the cavity to prevent fraying. Place a heavy object on top so it dries flat.

3.13 Apply Weight

Your book safe is done and ready to safeguard your valuables.

3.14 Book Safe

━ ━ ━ ━ ━ ━ ━ ━ ━ ━ ━ ━ ━ ━ ━

Securing your castle can be a long, time-consuming process. But if and when you first scan the horizon and notice the silhouettes of mounted savages riding toward you, you'll be glad that you made the effort.

The first step toward securing your castle is to give thoughtful consideration to the offensive and defensive measures that will be most effective against the expected threat. History shows that Mongols and Huns are particularly vexing threats, but with your valuables well hidden, a sturdy defensive wall protecting your home, and a finely made crossbow showing you mean business, you can sleep at least a bit more soundly.

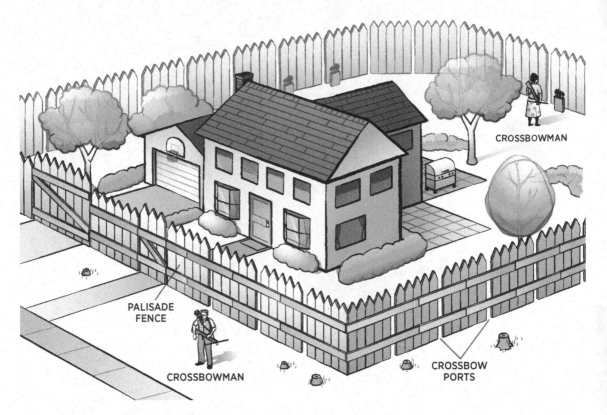

CROSSBOWMAN

PALISADE
FENCE

CROSSBOWMAN

CROSSBOW
PORTS

3.15 By adding a palisade wall and ports for crossbows, you can improve your odds against many attackers, including Huns and Mongols.

4

ALEXANDER THE GREAT AND THE GREEKS

On the shelves of my local public library I found a wonderful if old-fashioned and rather dusty book called *Alexander the Great: The Merging of East and West*. Written a hundred years ago by a history scholar named Benjamin Wheeler, it remains, to my eyes, one of the best accounts of Alexander's march of conquest from Macedonia east to the Levant into Persia, Asia Minor, and India. The chapter entitled "The Siege of Tyre" makes clear the need for everyone to have a plan to defend their castle.

Imagine yourself a resident of Tyre, an island city just off the coast of modern Lebanon. In 332 BCE, the Tyrians, a people of Phoenician descent with a large and powerful navy, were pretty much minding their own business when they found themselves under siege by the Macedonian Greek army of Alexander the Great, who was determined take the city.

4.1 Alexander

"The island upon which the city was built," reads a chapter of Wheeler's book,

> was separated from the mainland by a channel, near the shore shallow and swampy, but over by the fortress reaching a depth of 18 feet. Being without ships, Alexander proceeded to build a dam, or mole, across the channel by driving piles and filling in with earth and stones. The abandoned houses of old Tyre, situated on the mainland opposite the island, provided a convenient quarry, and the hills of Lebanon, hard by, furnished timber for the piles and the siege machinery.

Seeing what Alexander and his men were up to, the Tyrians turned to their powerful navy. Says Wheeler, the Tyrian ships, well manned with archers and slingers, swarmed around the dam builders, driving the laborers from their work. In response, Alexander ordered that barricades be built to shelter the workmen. His men built multistory towers, outfitted at each story with catapults and mechanical bows and protected by thick layers of animal hide against missile attack and firebrands. The towers held the Tyrian ships at bay, for a while.

But Alexander realized that he still could not complete the mole and attack the island city without a navy. Leaving his engineers to continue the work, he departed Tyre with a small contingent of bodyguards with the aim of assembling a fleet that could counter Tyre's navy.

Returning to the cities he had conquered in the years preceding the Tyre engagement, Alexander appropriated his first ships from Aradus, Byblus, and Sidon. Then, he added several more ships from the cities of Rhodes, Soloe, Lycia, and Mallus. Little by little, Alexander's navy grew, but it was still no match for the powerful Tyrian fleet.

Alexander's naval ambitions were suddenly and unexpectedly given an enormous boost when the king of Cyprus read the political writing on the wall and decided to ally himself with the winning side. He sent 120 ships from Cyprus to join the Macedonians. Alexander found himself suddenly possessed of a superb fleet 200 ships strong. From this time on, the siege of Tyre would become a different undertaking, and before long, Alexander set forth from his naval base in Sidon harbor.

A few hours' sailing brought the fleet off the northern harbor of Tyre, where it halted in full battle array. At first the Tyrians decided to meet Alexander's force head-on with their navy, but when they counted the ships, they were shocked. Surprised by Alexander's new alliance with the Cyprian navy, Tyre's admirals realized that they were outmatched. They returned to port.

Alexander pressed forward with renewed vigor. Over the previous months, Alexander enticed the very best engineers from all over the Mediterranean region to join his camp by offering high wages. The engineers immediately set to work building newer and more powerful siege engines.

Concurrently, Alexander's army continued to expand the mole, bringing it ever closer to the city walls. No longer were the Tyrians able to bombard the workers at will; under protection of the combined Macedonian and Cyprian fleet, the workmen were safe from attack by sea, and so the work progressed rapidly.

Soon, the Macedonian mole had extended far from shore until it was under the shadow of Tyre's massive eastern city walls. These walls were enormous, the stalwart protectors of a city universally considered unassailable. Carefully constructed of hewn stone set in cement, the tops of the walls soared 150 feet above the water's edge.

When at last the mole was complete, thousands of armed Macedonians swarmed the top and manned the siege weapons. Alexander and his troops began to ply the great, metal-weighted battering rams that swung out across the water-gap and thudded against the walls' solid masonry. The Tyrian defenders high up on the ramparts pitched down great boulders upon the Macedonians and their machines. Day by day, hour by hour, the battle became bloodier and more and more frantic. The Tyrian's spirit sank lower and lower.

The Tyrians had hoped in vain that their allies from unconquered cities in the Mediterranean, most prominently Carthage, would arrive to save them. But no rescuers ever appeared.

Once the fact that they were completely on their own had sunk in, the Tyrians made a last, desperate attempt to disrupt Alexander's attack. They split their navy into two fleets, one based in the north harbor, the other in the south. There they waited in relative safety. After a few days, Alexander's ships that had been patrolling the northern end of the mole left the area to visit the mainland northeast of the city to resupply with food and water. At the same time, the Tyrian observers on the southern wall could see that Alexander was absent from the area and the Tyrians decided the time was ripe for action. Manned with the hardest-pulling oarsmen and the best-armed fighters, 13 of their best ships—three quinqueremes, three quadriremes, seven triremes—left the harbors.

One fleet sailed north as quietly as possible, gliding into the sea in single file. When the moment was right, their captains swung the ships hard toward the east in battle formation. They were less than a half mile from their targets before they were noticed. By the time Alexander's navy took up the alarm, it was almost too late.

The Tyrians smashed into the unwary Macedonian fleet. They rammed and sunk the great five-banked galley belonging to the king of Cyprus, then turned and sank three more ships belonging to Alexander's allies.

They pressed their attack, driving other enemy ships against the rocky shore. Pulling alongside, they found that some enemy ships were empty because their crews had gone ashore for provisions. Gleefully, the Tyrians clambered over the sides of their ships to smash the empty warships to bits. The work of destruction, says Wheeler in his account, "went merrily on."

Despite being taken by surprise, Alexander reacted quickly. He commanded his quinquiremes and triremes to immediately push out to sea. Others were ordered to follow as fast as they could be manned. Unfortunately, Alexander and his relief forces were south of the mole and thus cut off from direct access to the action at the northern end of the city.

Alexander and his fleet rowed out to the west to circle the fortress and engage the Tyrian fleet as it returned to the north harbor. From atop the city walls, Tyrians observers saw Alexander's southern fleet pulling hard for the northern harbor. When they saw Alexander himself leading the counterattack, their exultation turned to dismay. They shouted and yelled, attempting to warn the Tyrian crews of their danger and call them to return, but they couldn't be heard above the din of the battle. "Too late," says Wheeler, "did the men see their danger."

Alexander caught them off the entrance to Tyre's northern harbor. Many of the Tyrian ships were shattered or sunk by ramming. A Tyrian quinquereme and quadrireme were captured outright in the mouth of the harbor. The Macedonian fleet closed their attack on the Tyrians who, tired and spent, could row no more. Within an hour the battle was over and the Tyrian navy vanquished.

Now, with nothing to fear from the Tyrian fleet, Alexander boldly pressed his attack upon the city walls. He ordered his catapult commanders to continually batter the walls. Over and over, great stones were flung against the walls. At first, there was little effect, but no wall can last forever against great force applied to a small area. Eventually a projectile broke through a section of the wall close to the southern harbor. Another direct hit enlarged the crack. The gap soon widened and Alexander's catapults reduced the wall to rubble.

Two shiploads of heavy infantry and skirmishers pushed their way through the gap. Led by Alexander himself, the frenzied Macedonians charged into the breach. The Greeks fought like lions, inspired by the presence of their king to extraordinary energy and viciousness.

The rest of Alexander's army simultaneously attacked the entrance of both harbors. He had outfitted several of his ships with battering rams, one of the few times in history that shipborne rams were recorded as being used. The rams did their work, and soon the harbors were in possession of Alexander's men and another large force of Macedonians raced into the city.

4.2 Siege of Tyre

The Tyrians, who had now given up defending the city wall, united for their last stand before the shrine of Agenor, just south of the northern harbor, and here the battle devolved into massacre.

"The main body of the Tyrians," wrote Wheeler, "deserted the wall when they saw it in the enemy's hands, but rallied opposite the Agenor shrine, and there faced the Macedonians. Against these Alexander advanced pursuing those who fled. Great was the slaughter wrought by those who had entered the city; for the Macedonians spared nothing in their wrath."

After making a sacrifice to the gods and Hercules, and dedicating the catapult and ram with which the walls had been battered, Alexander celebrated his victory with a grand military parade, naval review, and athletic tournament.

THE TORTOISE AND THE RAM

Why a section on how to build a battering ram in a book about *defending* your castle? Well, this goes under the general heading of *know thy enemy*. If you are familiar with the construction and operation of the ram, you'll be far better prepared to defeat it should you be on the receiving end. And who knows, you may need to go on the offensive at some point and this is a skill that will stand you in good stead.

At first blush, there doesn't seem to be much involved in understanding the battering ram. Cut down a tree, have a bunch of soldiers grab onto it, and run toward a door—that's about it, right? Well, that's true to some extent, but the story runs deeper.

Ancient armies used several types of battering rams. The earliest is the simple ram. Indeed, it is merely a wooden spar, hoisted to the shoulders of a group of warriors who repeatedly ram into things. But, simple as these rams were, they were large and powerful. Mark Antony, a famous Roman general of the first century CE, led his men in the use of a battering ram 800 feet long.

Marcus Vitruvious Pollio was a famous Roman military historian who wrote extensively about siege engines such as catapults and battering rams. He wrote that the first use of more complex battering rams occurred when the Carthaginian army battered down the walls of the city of Cadiz. A military engineer named Pephasamenos thought up the idea of adding flexible handles to a large wooden beam, providing the besiegers with a better way to hold and swing the ram. "Thus swinging it back and forth," wrote Vitruvious, "they leveled with heavy blows the walls of Cadiz."

But the men wielding the ram were still exposed to rocks, arrows, burning sand, and boiling water being rained down on them by a city's defenders situated on the walls above. Pephasamenos's coworker, Cetras of Chalcedonia, came up with the idea of adding a movable frame topped with thick leather hides to protect the ram-wielders while they banged against city walls. Cetras called his machine the tortoise because the hides reminded the soldiers of a tortoise's shell, plus this was one very slow-moving machine.

The protective shell provided the superstructure within which to create a more complex battering ram. This led to a truly brilliant insight on the part of Greek and Roman battering ram inventors. Instead of using human strength to attack the walls, these engineers built a ram suspension system that allowed the ram to be slung from ropes attached to the beams and rafters of the tortoise. In this way, the substantial weight of the log could be forcefully applied against the walls when swung back and forth by soldiers.

Although no wall could be constructed strong enough to withstand unlimited, well-targeted blows from within a tortoise, the wall-breaking power of a ram could be countered, at least to some extent, through clever engineering.

In 70 CE, the Roman army under Vespasian besieged the Jewish fortress at Masada. Vespasian ordered his army to build several rams. Josephus, the Jewish commander, came up with a clever plan.

According to his diary, Josephus

> gave orders to fill sacks with chaff, and to hang them down before that place where they saw the ram always battering, that the stroke might be turned aside, or that the place might feel less of the strokes by the yielding nature of the chaff. This contrivance very much delayed the attempts of the Romans, because, let them remove their engine to what part they pleased, those that were above it removed their sacks, and placed them over against the strokes it made, insomuch that the wall was no way hurt, and this by diversion of the strokes.

4.3 Machines Used for the Defense of Stone Walls

Charles Knight, Old England: A Pictorial Museum (1845)

Alexander's Tortoise

You may think that it's easy to build a battering ram. Well, you're right. As far as weapons of war go, you won't find one much simpler. Still, there are a few tricks that separate a really good battering ram builder from a mediocre one.

Note: Although this is a model, its dimensions and construction can be scaled up in case you need to take a real run at something.

MATERIALS

- ☐ (4) 1-inch × 1-inch square wooden dowels, 2 inches long: Wheel Mounts
- ☐ Wood glue
- ☐ (4) 1-inch × 1-inch square wooden dowels, 6½ inches long: Short Horizontal Frames
- ☐ (4) 1-inch × 1-inch square wooden dowels, 8½ inches long: Long Horizontal Frames
- ☐ (4) 1-inch × 1-inch square wooden dowels, 3½ inches long, one end on each piece cut at 45 degrees: Rafters
- ☐ (1) 1-inch × 1-inch square wooden dowels, 10½ inches long: Ridge
- ☐ (4) 1-inch × 1-inch square wooden dowels, 4 inches long: Vertical Frames
- ☐ (1) Piece of leather or other heavy fabric, 10 inches × 11 inches: Cover
- ☐ (6 to 10) #0-sized grommets
- ☐ (1) 1-inch-diameter round hardwood dowel, 11½ inches long: Ram
- ☐ (2) ⅜-inch round wooden dowels, 3 inches long: Ram Supports
- ☐ (8) Small screw eyes (#212 or similar)
- ☐ (1) 1-inch or ⅞-inch metal-tipped furniture leg glide, nail-on style: Ram Head
- ☐ ⅛-inch-diameter cord, 3 feet long
- ☐ (4) Axle pins for wooden wheels (Available in craft stores)
- ☐ (4) 2-inch wooden wheels (Available in craft stores)

TOOLS

☐ Electric drill with ⅜-inch and ¼-inch bits

☐ Grommet setting kit

☐ Hammer

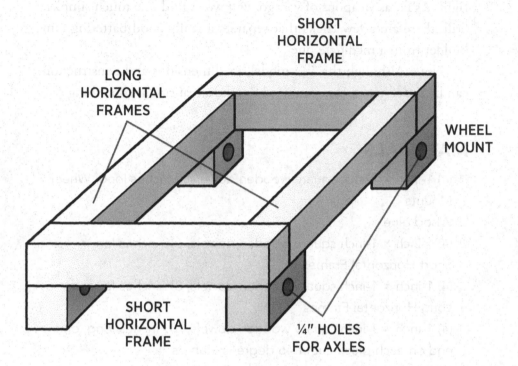

4.4 Lower Frame Assembly

1. Drill holes for the axles (usually ¼-inch, but check axle diameter before drilling) in the 2-inch 1 × 1 inch dowels for Wheel Mounts, as shown in **diagram 4.4**.

2. Using the wood glue, construct the lower frame of the Tortoise as shown in the diagram. The Short Horizontal Frame pieces are the 6½-inch dowels, and the longer sides are the 8½-inch dowels. Position the Wheel Mounts in each corner, glue in place, and let dry.

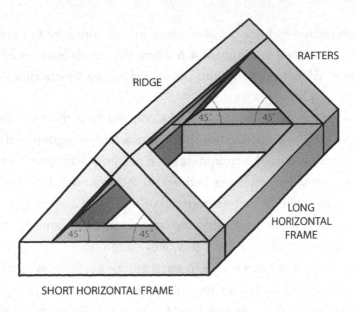

4.5 Upper Frame Assembly

3. Using glue, assemble the wooden pieces of the upper frame of the Tortoise. Just like the lower frame, the Short Horizontal Frame pieces are the 6½-inch dowels, and the longer sides are the 8½-inch dowels. The Rafters are made of the 3½-inch dowels with the 45-degree cuts placed on the frame pieces. The 10½-inch dowel fits in place at the top as a Ridge.

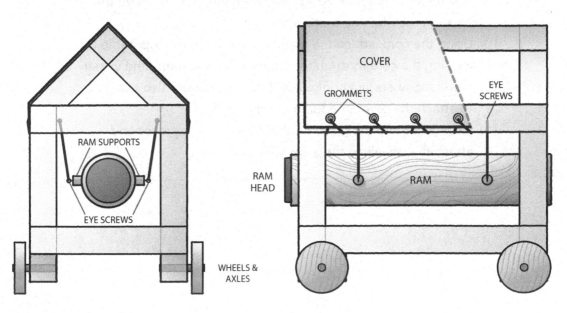

4.6 Ram Assembly

4. Place each 4-inch-long dowel piece in one corner of the lower frame as shown in **diagram 4.6**. Place the upper frame atop the pieces. Glue everything into place and let dry. Check glue label for drying times.

5. Attach grommets evenly spaced along the 10-inch-long sides of the leather piece as shown in the assembly diagram. Follow the directions on the grommet kit to securely add grommets.

6. Drill parallel ⅜-inch holes in the 11½-inch round oak dowel, approximately 2¾ inches from either end. Add a bit of glue and insert the ⅜-inch dowel into the hole so that the same length of wood extends on both sides of the large dowel.

7. Attach one small screw eye to each end of the ⅜-inch dowels. You may need to drill a small pilot hole to get the screw eye started depending on how hard the wood is. Next, attach the remaining four screw eyes to the upper frame, 2 inches from each end, as shown in **diagram 4.6**.

8. Hammer the metal-tipped furniture glide onto one end of the 1-inch dowel.

9. Attach the battering ram to the upper frame by tying the cord from each screw eye on the ⅜-inch dowel to the closest corresponding screw eye on the upper frame. Cut the cords so that the battering ram hangs approximately 1 inch above the lower frame when tied into place.

10. Using the cord, attach the leather covering to the top frame by looping the cord through the grommets and around the upper frame members in a continuous fashion. You can also use glue to attach the cover in addition to the cord.

11. Insert axles into the holes on the Wheel Mount pieces and attach the wheels to the axles with glue.

4.7 Alexander's Tortoise

Congratulations, your battering ram is ready to take on castle walls.

■■■■ - ■ - ■■ - ■ - ■ - ■■ - ■ - ■ - ■■ - ■ - ■■■

BATTERING RAM SCIENCE

Once the ram-wielding soldiers chose a likely spot for their assault, they drew close to the target area. They grabbed ropes that were attached to the swinging ram and pulled back. As they pulled, the ram moved upward, like the hands moving from six o'clock to nine o'clock on a clock face. On their leader's signal they let go and the ram crashed into the wall.

The amount of energy imparted on each blow of the ram is equal to the mass of the ram times the vertical height to which it was raised times a constant based on the acceleration of Earth's gravity. The equation looks like this:

Energy of blow = mass of ram × height the ram is raised over its rest position × gravitational constant

So, to add more energy on each ram blow, the Macedonian soldiers could either add weight to the ram or pull back harder, thereby raising the ram higher on the arcing path made by the swing of the rope.

SHIELDS UP

At some point in just about every episode of *Star Trek*, a Klaxon horn suddenly blares, officers on the bridge run around urgently, and the second in command shouts in a most alarmed tone "Shields up! Red alert!"

The "shields up" command as directed by Riker or Spock meant activating some sort of energy field that would prevent incoming missiles, torpedoes, and death rays from penetrating their ship's hull and harming the crew. That's a pretty futuristic shield and would be a great thing to actually have, if it weren't impossible according to the laws of physics.

Shields—and their close cousins, armor—have long been an important part of siege warfare and pitched infantry battles. In fact, "shields up" may be the oldest battle command in history. Before there were guns, bows and arrows, and even before swords were invented, there were shields. For over 2,000 years the shield was a vital piece of military equipment. Everyone, from the lowest peasant to the highest noble, would have used one. In many cultures the shield was the mark of a warrior, even more so than the sword or spear. The Roman historian Cornelius Tacitus wrote, "To lose one's shield is the basest of crimes." The shield carried great symbolic weight for Spartan soldiers. If they died in battle, their comrades carried their bodies back strapped to their shields. Leaving for war, Spartan mothers of the Classical period told their sons, "Come back with your shield or upon it." (Spartan mothers were not known for overindulging their children.)

According to the well-known British military historian Oliver Hogg, the shield was quite well known to our prehistoric ancestors. Taking cover is a natural human reaction to a threat. The earliest protohumans jumped in a hole or hid behind a boulder when danger arose. But hiding behind a stationary object prevented the cavemen from taking offensive action, so portable protection—a shield—soon appeared and provided far greater opportunities to its holder.

According to Hogg's chronology, the earliest shields were made of wood or hide and were fashioned in many sizes and shapes. Around 1000 BCE, the first metal shields, made of bronze, appear in the archaeological record. Bronze Age shields varied greatly, but a typical shield was circular with raised edges, measured about two feet in diameter, and was fitted with a hand grip in the center.

The shields of Alexander the Great's soldiers are particularly well known. Greek soldiers were known as hoplites because of the innovative interlocking shields they carried, called hoplons. A hoplon was made of bronze and wood often backed with leather. These were heavy shields weighing as much as 18 pounds. An important battle technique called the phalanx was based on the use of hoplons.

In the phalanx battle technique, Greek commanders arranged their men into densely packed squares of soldiers, each square fronted by a tight line of shields with long spears poking through. Once so arranged, the phalanx would march into battle rather slowly but nearly inexorably, pushing or spearing any enemies who dared to get too close.

Despite the substantial weight of the hoplon, it was still vulnerable to attack by spear and sword. In phalanx against phalanx battles, the first line of hoplite soldiers would often be skewered or fall down early on, requiring those in the following ranks to move forward to fill gaps. The deepest and most disciplined phalanx won the day.

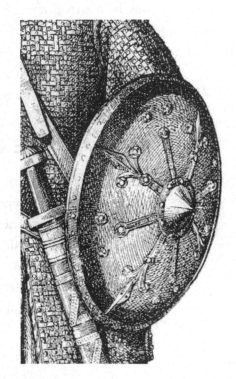

4.8 Round Shield

L'Art Pour Tours, Encyclopedie de l'Art Indstriel et Decratif

Over time, shields evolved in size, shape, and construction. The Roman legionaries, for example, were outfitted with scuta, which were large shields bent into a semicylindrical shape. This type of shield covered the holder entirely, protecting him from swords and arrows, at least as long as it stayed in one piece.

Horse-mounted medieval knights carried large kite-shaped shields rounded at the top and V-shaped at the lower edge. Historians say this shape was popular among both the Normans and the Saxons at the Battle of Hastings in 1066.

As time passed, the handheld shield became less popular among foot soldiers. It was heavy to carry and cumbersome to use. Shields shrank in size and eventually were eliminated in favor of mobility and the ability to use both hands to handle larger weapons like pikes and broadswords.

4.9 Kite Shaped Shield
The Century Dictionary *(1883)*

KEVLAR: THE BULLETPROOF FABRIC

Here is a counterfactual question to consider: how would the history of the world be different if those shields had not been so heavy and, even better, had been nearly impervious to arrows and spears?

DuPont Corporation makes an incredibly tough fabric called Kevlar that is five times as strong as steel on an equal-weight basis. Because it looks and feels similar to heavyweight cotton canvas, it is used to make bulletproof clothing and panels.

4.10 Kevlar Fabric

While it's a bit pricey, it's not incredibly so, and Kevlar is easily worked with inexpensive tools. It can be hemmed and joined on a normal sewing machine.

4.11 Hemming Kevlar

Kevlar was invented in the mid-1960s by Stephanie Kwolek, a researcher at DuPont. She was experimenting with compounds called polymers, which are long chains of molecules. She came across one formula that produced a breathtakingly strong fiber. Company chemists refined the fiber, which was soon introduced to the automotive industry as a reinforcement for rubber tires. But another use, as the basis for bullet-resistant clothing, is where the material really shines.

In the 1970s, a pair of civilian US Army scientists, Lester Shubin and Nicholas Montanarelli, first saw Kevlar's potential as a lightweight bulletproof fabric. To test out the idea, they covered a telephone book with a few layers of Kevlar and shot it with a .38-caliber handgun.

The bullets bounced off.

Encouraged, they developed additional tests to determine just how tough this stuff was.

Kevlar stood up to the challenge in test after test. The results were widely reported, and soon police officers across the country were wearing Kevlar vests. According to DuPont, its product has saved the lives of more than 3,000 people.

The High-Tech Backpack Shield

Note the Following Carefully

You don't need me to tell you that all knives, spears, and guns are dangerous. So, don't think for a moment that this project will keep you safe if bullets and arrows fly. While it's relatively easy to construct a bullet-resistant shield that will fit into a typical backpack, and while this insert could conceivably provide a measure of personal protection from medium-caliber handgun ammunition, *don't count on it!* At best, the shield described here is "bullet and projectile resistant" and definitely not "bulletproof." Many, many things affect the penetration power of a projectile, and there is no guarantee whatsoever that the panel that this project describes will stop any projectile. Build and use at your own risk.

Before you begin, you'll need some familiarity with the materials to use. There are several different types of Kevlar fabric, differing in a property called elastic modulus. The type you want for a bullet-resistant shield is the low elastic modulus type designated as Kevlar 29. The more common Kevlar 49 is designed for making reinforced composites but is not used for protection from ballistic attack. Look for fabric suppliers on eBay or use the following Internet search term: "Kevlar 29 fabric." Not all cutting shears will handle Kevlar fabric. Several manufacturers make affordable scissors specifically for cutting Kevlar fabric. Search online for "Kevlar shears."

TOOLS AND MATERIALS

☐ Safety glasses
☐ Rubber gloves
☐ Tarp for workspace
☐ (2) yards Kevlar 29 fabric
☐ High-strength scissors
☐ Disposable mixing bowl
☐ Stir stick
☐ 1 quart all-purpose fiberglass resin with hardener (Fiberglass resin is usually available in hardware, auto parts, and home stores)
☐ Wax paper
☐ Resin spreader (spatula)

1. Put on safety glasses and rubber gloves and cover your workspace with the tarp. Make sure that you are working in a well-ventilated area.
2. Cut the Kevlar sheets into 15 identical panels of the general width and height of your backpack, generally about 10 inches by 12 inches. With the right scissors, Kevlar cuts fairly easily, although is it very prone to fraying. To reduce the amount of fraying, avoid handling the cut panels.
3. Mix up 8 ounces of resin and hardener in the disposable mixing bowl. (Mix additional resin as needed.) The resin is pretty messy stuff, so avoid getting it on your clothes, body, or anything you care about. Carefully follow all directions on the resin container.

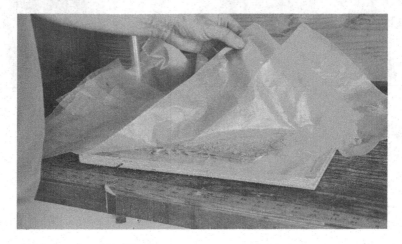

4.12 Laying Up Resin

4. Cover the work surface with wax paper. Place the first panel on the paper and use the resin spreader to saturate the Kevlar fabric with resin. Place another panel on top of the first panel and again saturate it with resin using the resin spreader. Continue adding Kevlar panels and resin until the panel is the desired thickness. Thicker panels, made from about 12 or more sheets, will provide a higher likelihood of actually stopping a bullet. When you have the desired number of layers, place a piece of wax paper on the top layer and spread about 20 pounds of weight on top.

   The resin, once the hardener is added, solidifies quickly. Place the layered Kevlar in a well-ventilated place to dry.

4.13 Fiberglass/Kevlar Panel

USING YOUR SHIELD

Hopefully, you'll never have to use your high-tech shield to ward off threatening Macedonian armies, Mongol hordes, or Viking warriors. But if the chips are down, it just might come in handy. Using your shield is pretty straightforward. Place the Kevlar shield in your backpack, on the side away from your body. In case of big trouble, hold the shield between yourself and the threat, using it to protect your head and upper body. Obviously, a backpack is not large enough to cover all vulnerable areas, so as soon as possible, take flight or take shelter.

TESTING RESULTS

Using the directions outlined above, I made a shield measuring approximately 12 inches by 14 inches that consisted of 12 pieces of Kevlar 29 laminated with fiberglass resin and hardener. With the help of a nationally recognized ballistics expert, I put the shield to the test.

4.14 Test Equipment

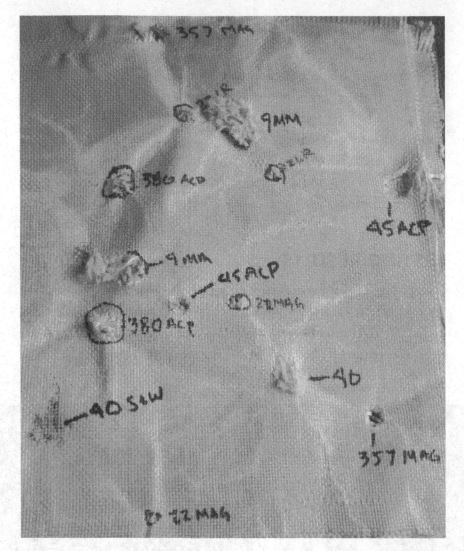

4.15 Results

The multilayered Kevlar shield was placed inside a backpack that was mounted on a stand at the back end of an outdoor shooting range. Several handguns of various calibers were fired at the backpack from a distance of 7 yards. Several bullets of each type were fired, and the results were recorded and photographed. The velocity of each shot was recorded using a pair of shooting chronographs. While our test does not have the rigorous controls and staging that a government-certified test for commercial products would require, we found our results to be consistent. The outcome of the test was mighty impressive:

Caliber	Handgun	Bullet Type	Result
.22 caliber	Sturm, Ruger Mark II Model 22/45 Target; 5½-inch barrel	Remington .22 caliber, high-velocity, 36 grain bullet	*Stopped*
.22 caliber	Colt Trooper Mk III, 6-inch barrel	Winchester .22 caliber magnum, high-velocity load, 40 grain, full metal jacket	The bullet penetrated the Kevlar shield, the backpack, and the wooden stand.
.380 ACP	Smith and Wesson Sigma SW380, 3-inch barrel	Winchester 380 ACP, 95 grain, full metal jacket	*Stopped*
9mm	Browning Hi-Power, 4⅝-inch barrel	Remington 9 millimeter, 147 grain Black Talon hollow point	*Stopped*
.40 S&W	Smith and Wesson Model 4006, 4-inch barrel	Winchester .40 S&W 165 grain, full metal jacket	*Stopped*
45 ACP	Colt Model 1911 Military, 5-inch barrel	Federal Cartridge .45 caliber M1911, 240 grain, full metal jacket	*Stopped*
.357 Magnum	Colt Python 6-inch barrel	Winchester .357 Magnum, 145 grain, Silvertip	*Stopped*
.44 Magnum	Smith & Wesson Model 29-2, 6-inch barrel	Hornady .44 Magnum, 240 grain, soft point	*Stopped*

In our test, the Kevlar backpack shield successfully stopped all bullets fired from the very large and intimidating Winchester .357 magnum and the uber-powerful Remington .44 magnum (famous in the *Dirty Harry* movies of the 1970s). The only handgun bullet it could not stop was the full metal-jacketed, high-velocity .22-caliber round.

4.16 Recovered Bullets

At first look, the fact that the small diameter .22-caliber magnum was the only bullet to penetrate all layers of Kevlar may seem surprising. But this bullet has a small diameter and was fully clad in metal, which kept the cross section small even after it hit the shield. Its energy was confined to a small area, and the bullet crossed relatively few of the fibers in the Kevlar fabric.

Although the test was carefully made, there are loads of disclaimers and modifiers that I must share. First, the size of the bullet's powder load, the jacketing on the bullet, the material from which the bullet is made, and the distance from which the weapon is fired all affect the power and penetration of a bullet. In the limited testing we carried out on this project, the shield stopped everything except the .22 magnum. However, it is a near certainty that under other conditions (or perhaps even under the same conditions), the shield will not stop these bullets every time.

It is important to note that the shield will *almost certainly not stop* ammunition fired from any rifle. Rifle bullets contain far more energy and will go through the shield as if it were made of cardboard. Further, even if the panel does stop a bullet and prevent it from entering a human body, the force of the bullet may be transmitted through the backpack causing potentially fatal blunt trauma injuries.

Still, the Kevlar backpack shield did appear effective at stopping most handgun-fired ammunition, and when bullets fly that is at least a bit better than nothing.

5

TAMERLANE AND THE TATARS

There is an old and often-told story that early in his rise to power, the powerful Asian chieftain known as Tamerlane was being routed in battle by a powerful enemy. With opposing soldiers all around him, he hid in a deserted building. As he concealed himself, miserable and desperate to escape, he saw on the ground before him an ant carrying a large seed of grain. Mesmerized, he watched this ant try to carry the grain, which was much larger than the ant, up and over a wall. Repeatedly the ant fell back unable to surmount the wall. But undeterred, the ant would load up and begin the climb over and over again.

Tamerlane began to count the attempts. Ninety-nine times the little ant attempted the climb and on each occasion she fell. On the hundredth try, she was able to push the piece of grain over the top. Tamerlane was inspired by this display of perseverance. He came out of hiding and rallied his forces, eventually turning the enemy and chasing them away. He learned an important lesson from that ant about determination, willpower, and tenaciousness, and it changed his life. Or so the story goes.

The story of Tamerlane and the ant is a cute one. But it's apocryphal, and it seems much out of character given the bloodthirsty and cruel nature of the man. There was absolutely nothing cute and cuddly about Tamerlane.

Of all the invaders who swept in from the steppes of Central Asia in the Middle Ages, Tamerlane (sometimes written Timur-lane or Timur) and his horde of assorted Central Asian peoples (somewhat incorrectly called Tatars by most Western writers) are certainly among those you would least want to see massing outside the gates of your city.

Tamerlane was born near the city of Tashkent, the modern-day capital of Uzbekistan. He claimed to be a descendant of Genghis Khan. Whether he was a direct descendant or, more likely, descended from one of the members of the khan's court is impossible to determine. But certainly he acted the part.

Militarily, his achievements are on par with those of Genghis and Alexander. His army of horsemen, riding westward out of the steppe in the late 14th century, brought fear and consternation to Europe on a scale that hadn't been known since Genghis Khan and his sons terrorized Hungary and Russia in the 1240s. Tamerlane defeated some of most powerful armies of the time including other Mongol hordes, the Ottoman Empire, and even the fierce slave-warriors of Egypt known as the Mamluks. He ranks as one of history's most ambitious men, and at his peak his empire ranged on the modern map from India to Turkey, and from the Aral Sea to the Persian Gulf.

And if even just some of the stories like the ones that follow are true, Tamerlane ranks as one of history's cruelest men as well.

Like Genghis, Tamerlane was born into a family of poor nomadic herdsmen. Early on, his ambition, courage, and military skill were apparent to those around him. Via a combination of skill, intrigue, and luck, he quickly rose to become the leader of a great army. His most famous victory, over the Ottoman Empire, led to the capture of

5.1 *Sultan Bayezid Imprisoned by Timur*

Stanisław Chlebowski (1878),
Lviv National Art Gallery

Sultan Bayazid I, known to his subjects as "the Thunderbolt." According-ing to a biographer, Ahmed ibn Arabshah, who lived at the same time as Tamerlane, Tamerlane humiliated the sultan, keeping him in chains within an iron cage and feeding him like a dog with scraps of food. Other witnesses wrote that they saw Tamerlane use Bayazid like a stool to stand upon when he mounted his horse.

> Black are his colours, black pavilion;
> His spear, his shield, his horse, his armour, plumes,
> And jetty feathers, menace death and hell;
> Without respect of sex, degree, or age,
> He razeth all his foes with fire and sword.
>
> —*Tamburlaine the Great, Part 1*,
> Christopher Marlowe

He was no less cruel to common people. There are many accounts of him issuing orders to kill unarmed women and children, although apparently he did give his victims an out of sorts. According to historian Aneas Sylvius Piccolomini, who was later chosen Pope Pius II, when Tamerlane laid siege to a city he had a white tent pitched for himself on the first day, a red tent on the second, and a black tent on the third day. All of those who surrendered while he was sitting in the white tent were allowed to live. But if the besieged waited until the red tent was up, then Tamerlane decreed that he would kill all the heads of the city's households, although he would spare the women and children. And, when the black tent went up and stayed up, there was no mercy: all would die and the city would be razed and everything reduced to ashes.

In Hungary in 1396, Ottoman Sultan Bayazid I captured a literate German soldier named Johann Hans Schiltberger in a battle with Christian knights. A few years later, in 1402, when Tamerlane's Tatar army defeated Bayazid, Schiltberger was made a personal slave to Tamerlane. He escaped to freedom in 1427 and wrote a book called *Travels and Bondage*, from which this is excerpted.

> Tamerlane went into a kingdom called Isfahan and required it to surrender. They gave themselves up, and went to him with their wives and children. He received them graciously and occupied the city with six thousand of his people. But when Tamerlane and his main army marched away, the inhabitants closed all the gates and killed all six thousand occupiers.
>
> When Tamerlane found this out, he returned to the city and besieged it and entered it. He assembled all the citizens, and ordered all those over fourteen years to be beheaded. He ordered that the heads be constructed into a tower in the centre of the city. He ordered the women and children to be taken to a plain outside the city and ordered his cavalry to trample them to death. Then he set fire to the city.

In 1400, Tamerlane invaded Syria, where he massacred hundreds of thousands of people and destroyed Damascus, then one of the richest and most beautiful cities on Earth. Next, he marched on Baghdad, where Tamerlane ordered each of his soldiers to bring him the head of a dead Babylonian.

This they did and soon a pyramid of 90,000 human heads was erected on the ruins of the city. Evidently Tamerlane had an odd fascination with skull pyramids. And this act, as violent and evil as it was, was not particularly notable given his long and bloody career. In another instance he ordered 4,000 soldiers along with their horses to be thrown into the moat of a city that he had conquered, and they were all buried alive. In another, he is said to have ordered 2,000 prisoners to be stacked one upon another while still alive and covered with cement to construct yet more skull pyramids.

DEFENDING THE CASTLE DOOR

The front door to your house may look big and strong, but is it really? There are many ways to get through a seemingly solid wall or door. If the Tatars can figure out a way to apply some force, well, a lot of force upon a small area, most doors won't last for long. A warp, a bend, a hole, and then it's done. Once the integrity of the door is breached, whatever security it provided is gone.

Getting through a locked door without a key is what your local police would call forcible entry. There are several forced entry techniques. First, there are the exotic methods. Chemicals can be applied to disintegrate or dissolve key components. Typically, these methods are directed against specific vulnerabilities such as lock parts or hinges. Shock methods involve using explosives to puncture plate or bulkheads, or to remove vulnerable locks and hinges.

Sanctioned civil authorities such as police and firefighters possess an array of modern tools designed to enter a space by the application of force. There's nothing stealthy or delicate about the use of rams and lever tools to defeat a door. Forcible entry is loud, difficult, and time-consuming, but in the end, it almost always works.

The standard tool in the firefighter's arsenal of entry tools is called a Halligan bar. This is a three-pronged, forged-steel crowbar designed to gain purchase in the gaps between doors and doorjambs. Once in place against a door, the Halligan allows the rescuer (at least one hopes the user is a rescuer) to apply a tremendous amount of splitting force by means of leverage. An adept user of a Halligan bar can get through nearly any door, window, crack, or wall crevice.

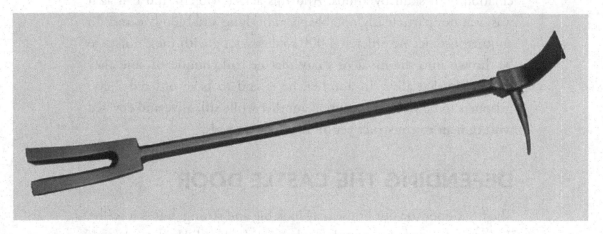

5.2 Halligan Bar

So, given enough time and effort, all castles can be breached. The doors on houses and most commercial establishments, short of high-security bank vaults, are easily rammed, kicked, or destroyed using levers. Modern wood doors are hollow and often contain windows or other weak spots. Although the locking bolt that secures the door from unwanted intruders is often made of steel, the frame into which it is set is typically softwood. It doesn't take too much to get through such lightly constructed barriers.

But still, historical invaders such as Tatars, as well as modern untrained attackers, can be turned away by solid door design and construction. Such doors are nearly invulnerable to the invader who lacks special entry tools or the training to use them.

The Tatar-Proofed Castle Door

Most people will find commercially made home security doors adequate for their needs. Doors constructed with 14- and 16-gauge steel, coupled with reinforced door jambs, hinges, and locks, are adequate for most real-world situations. A well-designed security door will resist just about everything short of an expertly wielded Halligan tool, or battering ram.

It is certainly possible to go beyond that, however. An attack-resistant door can be rigged up from a layered combination of steel plates, high-strength fasteners, and energy-absorbing cushioning material. The thick steel facing is strong and hard enough to fend off attacks from edged weapons, and the energy-absorbing polymer inside makes it difficult for ram attacks to succeed. The energy-absorbing material is a synthetic viscoelastic urethane polymer often used as a shock absorber and vibration damper.

Now, before you run off to the steelyard and start buying steel, keep in mind that steel plate is expensive and heavy, and it is very difficult to work without special tools and equipment. You definitely need at least a gin pole (see chapter 3) to move it, and a forklift would be preferable. And the larger the door, the more difficult the construction becomes. Based on my experience, a full-sized, fully reinforced, truly Mongol-proof steel door is an undertaking well beyond most nonprofessional metal workers. Seven-gauge steel plate is a full $3/16$-inch thick and weighs 7½ pounds per square foot. Since the average-sized front door measures about 3 feet by 7 feet, the steel alone weighs nearly 160 pounds. And that's before adding cladding, wood backing, rivets, and energy absorbers. By the time you're done, you've got a 200-pound door! Unless you're preparing for trouble involving rocket attacks and bulldozers, this is likely more than what's needed to protect yourself. Nonetheless, it's fascinating to know how such doors are constructed.

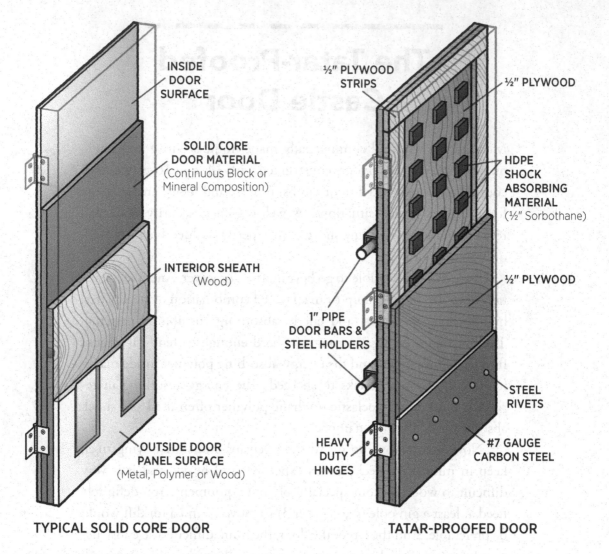

INSIDE DOOR SURFACE

SOLID CORE DOOR MATERIAL
(Continuous Block or Mineral Composition)

INTERIOR SHEATH
(Wood)

OUTSIDE DOOR PANEL SURFACE
(Metal, Polymer or Wood)

½" PLYWOOD STRIPS

½" PLYWOOD

HDPE SHOCK ABSORBING MATERIAL
(½" Sorbothane)

½" PLYWOOD

1" PIPE DOOR BARS & STEEL HOLDERS

HEAVY DUTY HINGES

STEEL RIVETS

#7 GAUGE CARBON STEEL

TYPICAL SOLID CORE DOOR

TATAR-PROOFED DOOR

5.3 Door Construction

MATERIALS

Note: Remove sharp burrs from metal edges with the flat file before handling. Build, move, install, and use this door at your own risk.

☐ (1) 7-gauge steel sheet, precut to the dimensions of the required opening, minus the width of the sheathing and the minimum door gap required to ensure the door opens smoothly

☐ Heavy-duty sawhorses

☐ (1) piece ½-inch-thick plywood, same size as 7-gauge steel sheet: Top Plywood Sheet

☐ (1) piece ⅜-inch-thick plywood, same size as 7-gauge steel sheet: Bottom Plywood Sheet

☐ (2) pieces of 1½-inch-wide, ⅜-inch-thick plywood, cut to the same length as the long edge of the 7-gauge steel sheet: Longer Inside Frame

☐ (2) pieces of 1½- inch-wide, ½-inch-thick plywood, same length as the short edge of the 7-gauge steel sheet, less 3 inches: Shorter Inside Frame

☐ All-purpose glue

☐ (1) 4-inch × 4-inch × ½-inch-thick sheet of Sorbothane or other shock-dampening material, cut in 16 1-inch × 1-inch squares (Enter "sorbothane" into an Internet search engine to find suppliers): Sorbothane Squares

☐ (24 to 36) Steel blind rivets, ³⁄₁₆-inch-diameter, each approximately 1½ inches long

☐ (24 to 36) Steel washers or ⅛-inch-diameter rivets

☐ (2) strips of 16-gauge sheet metal, 4¾ inches wide by the length of the longer end of the door: Longer Door Jacketing

☐ (2) Strips of 16-gauge sheet metal, 4¾ inches wide by the width of the shorter end of the door less 3 inches: Shorter Door Jacketing

☐ (4) Heavy-duty hinges with bolts and nuts

☐ (2) 3-inch steel C-channels, cut to height of opening: Side Door Frames

☐ (2) 3-inch steel C-channels, cut to width of opening: Top and Bottom Door Frames

☐ (1) ½-inch × ½-inch angle iron, same length as the long edge of the door: Door Stop

☐ (16) ⅜-inch-diameter lag screws, or other heavy-duty wall anchor: Lag Screws

☐ (2) ⅛-inch-thick steel strips, ¾ inch wide, 8 inches long: Bar Bracket

☐ (2) 1-inch-diameter steel pipes, 8 inches longer than the door opening width: Bars

TOOLS

- ☐ Measuring tape
- ☐ Straightedge
- ☐ Flat file
- ☐ Electric drill with full set of bits
- ☐ Rivet gun with 3/16-inch riveting head
- ☐ Brake for bending metal or a set of metal and hard rubber hammers
- ☐ Wrench set
- ☐ Welder, welding rod or wire, fixtures, and protective equipment

DIRECTIONS

1. Place the 7-gauge sheet on heavy-duty sawhorses. It will be heavy, so get assistance. Place the Top Plywood Sheet atop the 7-gauge sheet and align the edges.

5.4 Strips and Pads

2. Place the longer 2 pieces of 1½-inch-wide, ½-inch-thick plywood along the longer edges of the plywood sheet, and the shorter 2 pieces of 1½-inch-wide, ½-inch-thick plywood along the shorter edge of the bottom plywood sheet. Glue into place. Next, evenly space the Sorbothane pads onto the plywood sheet and glue in place. Place the Bottom Plywood Sheet atop the Sorbothane squares and door frame plywood pieces. Glue into place.

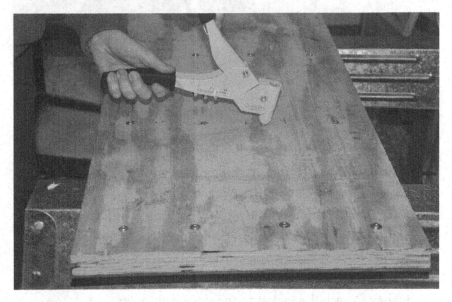

5.5 Riveting

3. Place the steel, plywood, and Sorbothane assembly on sawhorses. Drill 24 to 36 regularly spaced holes for the ³⁄₁₆-inch rivets through the sandwiched steel, plywood, and Sorbothane assembly. Back up the rivets on the plywood side of door with washers. Insert rivets following the instructions that came with your riveting tool. Be aware that using steel rivets with a hand riveter is hard work!

5.6 Bending Door Jacket

4. With the brake or your set of hammers, bend the longer and shorter strips of 16-gauge sheet metal into door jackets by bending U-shapes with the bottom of the U just wide enough to accommodate the sandwiched 7-gauge sheet, top plywood sheet, bottom plywood sheet, and inside door frame pieces, with about 1½ inches of overlap. If you used ⅜-plywood and 7-gauge steel sheet, the door thickness would be about 1½ inches. Attach the jacket to the door with rivets.

5.7 Jacket

5. Using the electric drill and wrench set, attach hinges to one end of the riveted door assembly.

5.8 Welding Door Frame

6. Weld the door frame from the 3-inch steel C-channels, with the open end of the C facing outward. Take care to align the channels carefully prior to welding. Use welding fixtures to maintain right angles.

5.9 Hinge

7. Using the electric drill, drill holes for the hinges through the door and the door frame's C-channels. Use a wrench set to attach the hinges on the riveted door assembly to the C-channel door frame with bolts.

5.10 Door Stop

8. Weld the ½-inch × ½-inch angle iron Door Stop to the inside edge of the long end of the door frame, opposite the hinges and oriented such that the stop prevents the door from opening outward.

9. At this point, you have a very strong and extremely heavy reinforced door and door frame. The next step is to attach it to your castle wall. Remove stone blocks or bricks to form a rough opening just slightly larger than the door frame. Move the door and door frame into place. Carefully align and level the door and make sure it opens and closes smoothly. Once you are satisfied, drill holes and then insert lag screws through the wings of the C-channels into the adjacent solid wall. Use stone or brick and mortar to finish the wall.

5.11 Door and Door Frame

10. Choose heavy-duty lock hardware to complete your door. Because this door is designed to swing in, I recommend you weld two sets of ⅛-inch-thick, ¾-inch-wide, 8-inch-long steel strips into bar brackets on the steel door frame as shown in the diagram. To lock the door, slide the two steel pipes through the bar brackets.

5.12 Bar Brackets

Congratulations! You now have a door that will stand up to nearly anything.

THE OTTOMANS RETURN

After Tamerlane imprisoned and eventually killed Beyezid I in 1402, the Ottomans were left without a ruler. Empires, just like nature, abhor a vacuum, and for the next decade Bayezid's sons fought over succession as the empire fell into chaos and disorder.

By 1413, Tamerlane was seven years dead. (He died of disease and old age while marching upon the Chinese.) Tamerlane's successors were not nearly as capable, and the empire he built crumbled.

As the Tatars retreated into the Central Asian steppes, the Ottomans rebuilt their empire and Bayezid's son Mehmet clawed his way to the top by pushing aside and then killing his brothers, and emerged as the new sultan. An able diplomat, Mehmet spent the next eight years reuniting his father's empire.

The Ottoman sultans who followed Mehmet—Murat II and Mehmet II—were no diplomats. They were fighters who threw themselves into a nearly constant state of belligerency with the European nations to their west. Murat II fought against the Hungarians and lost, and Mehmet II battled the Byzantine Greeks and won. Still referred to as Fatih (the Conqueror) by the Turks, Mehmet II defeated the Byzantine army and took the Queen of Cities, Constantinople. Mehmet renamed it Istanbul and made it his capital.

After Mehmet II, his son Beyazid II came to power and was an excellent diplomat. Under Beyazid II, the Ottoman economy flourished and wealth flowed into the empire. But at the same time, he worried the Europeans by carefully applying political and military pressure on the eastern Mediterranean. Nowhere was the worry about Turkish expansion and aggression as great as in Italy.

At the turn of the 16th century, Beyazid II ordered his able admiral, Kemal Reis, to sail against the Venetian fleet. The Venetians and the Ottomans were fighting over the islands in the Aegean, Ionian, and Adriatic Seas. In the two Battles of Zonchio, which took place off the southwestern coast of Greece, the Ottoman navy soundly defeated the Venetians.

Worried that another Tamerlane was setting up camp on their eastern doorstep, many Italian city-states, including Florence, the

Vatican, Milan, and Venice, began building up their defenses. One of the most notable efforts was put forward by Cesare Borgia, son of Pope Alexander VI (popes had different rules in those days, evidently) and commander of the large and powerful papal army.

In 1502, Borgia's army had subjugated nearly all of central Italy. But despite this, he was still concerned about his ability to cling to his power. Borgia, a notoriously shady character, had made many enemies. But he was extremely intelligent, and in a masterstroke of genius, Borgia hired Leonardo da Vinci as his chief military engineer. The brilliant Leonardo and Borgia traveled up and down the Italian coast, inspecting fortifications and designing improvements to Borgia's defenses.

DA VINCI'S NOTEBOOKS

Leonardo was born in 1452, about 50 years after Tamerlane's brutal campaigns. He was one year old when Mehmet the Conqueror captured Constantinople. Such events were not easily forgotten. Like modern baby boomers who spent their childhoods worrying about nuclear war with the Soviet Union, Italian citizens like Leonardo spent most of their lives believing themselves to be under threat from the Muslims.

Ottoman cannons were big enough to besiege a city from miles away, and the loss of Constantinople, whose walls were so high and strong that the city was deemed unconquerable, filled every kingdom in Christian Europe with worry.

While working for Borgia, Leonardo carried a notebook attached to his belt. In it, he would sketch ideas for machines that he believed would make his employer's lands impregnable—not just from the Ottomans, but from anyone.

Leonardo's notebooks are treasure troves of interesting designs, often containing plans and ideas for inventions that would come to pass hundreds of years later. Originally, Leonardo's notebooks were filled with sketches of people and objects of nature, probably as a way for him to remember something he had seen so he could reproduce it later as a part of a painting. But the notebooks soon came to contain words as well as pictures and became more or less permanent records of his ideas and thoughts about invention.

Leonardo filled many, many notebooks. When he died in 1519, he was a renowned artist and his notebooks were thought to be worth keeping. His friend Francesco Melzi inherited the notebooks. Melzi held his friend's works in great esteem and kept them safely in his possession. But after Melzi died, the notebooks were given away or sold, and many wound up in the hands of a Spanish artist named Pompeo Leoni.

In an ill-considered move, Leoni cut up the original notebooks and rearranged the material by subject. Doing so disrupted the chronology of notebooks, so we are left without an accurate time line as to when Leonardo devised many of his ideas. Each rearranged book of Leonardo's notes is known as a codex. There are 10 Codices in all, and each has been given a name; for example, the Arundel Codex, the Codex Forster, and the Madrid Codex. The most important codex, and the one with his military ideas, is the Codex Atlanticus, which is kept in the Ambrosiana Library in Milan, Italy.

It is in the Codex Atlanticus that you can find da Vinci's plans for a giant crossbow, a sketch for a siege engine large enough to knock down city walls, and designs for castles and fortifications. Perhaps the most interesting and practical plan is his sketch of an elegant, compact, yet high-powered tension catapult.

Da Vinci's Catapult

Possessing artillery gives you a big advantage when facing down a threat. A good defense only goes so far. To repel an invader, it's important to take the fight to your enemy as well. That's where artillery comes in. Leonardo understood that, and that's why Leonardo jotted down a bunch of ideas for improved, spring-powered artillery devices or catapults during his time in Milan before he began his employment with Borgia. Doubtlessly, he kept his best ideas with him as he advised his new boss on the best practices in military engineering.

Da Vinci's upgraded and reimagined catapult is a tension-powered hurling machine. It derives its power from multiple bent wooden springs placed on either side of the machine's central axis. The springs spin the machine's main axle, which in turn hurls a projectile placed in a shooting pouch attached to the main throwing arm.

Although it is clearly related to the crossbow and the ballista of ancient Greece and Rome, da Vinci's catapult is more refined and more powerful. By using several tension springs simultaneously, this device has considerable power yet has only a modest spatial footprint.

Leonardo's bright idea was the way he improved the efficiency and range of earlier designs. By Leonardo's time, catapults were losing ground to gunpowder weapons, although hurling machines were still used in some circumstances. Leonardo's design uses a particular type of energy storage device called a leaf spring. When bent, the fibers in the leaf spring store a great deal of energy. When the spring is released—that is, allowed to spring back to its normal shape—it releases a great amount of energy, which is used to hurl the projectile via the throwing arm.

This project is based on Leonardo's catapult design found in the Codex Atlanticus, although it's been redesigned a bit to make it easier to build. Like Leonardo's design, this too is compact and powerful. And while it's just a model and won't provide you with much protection from invading Mongols, it is great fun to build and provides good woodworking practice. And, it could be scaled up in case the need for the real thing comes up.

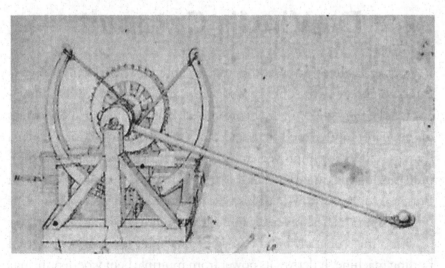

5.13 Leonardo's Catapult

MATERIALS

- ☐ (6) ¾-inch-diameter hardwood dowels, 8 inches long: Axle Pieces
- ☐ (1) ¾-inch-diameter hardwood dowel, 10½ inches long: Axle Center
- ☐ Masking tape
- ☐ Wood glue
- ☐ (6) ¼-inch-diameter wooden dowels, 2 inches long
- ☐ (4) Small screw eyes, ³/₁₆-inch-diameter eye or larger
- ☐ (1) ½-inch-diameter hardwood dowel, 13½ inches long: Throwing Arm
- ☐ (1) Small nail
- ☐ (2) 1-inch × 6-inch (nominal) pine boards, 15 inches long: Sides
- ☐ (1) 9½-inch × 15-inch × ¾-inch piece of plywood: Base
- ☐ (8) 1⅝-inch deck screws
- ☐ (7) 2½-inch deck screws
- ☐ (2) 2-inch × 4-inch (nominal) pine boards, 9½ inches long: Ends
- ☐ (1) Pencil eraser, 2⁹/₁₆-inch × 1-inch × ⁷/₁₆-inch (nominal)
- ☐ (4) 1¼-inch × ¼-inch × 20-inch oak strips (Available in trim section of a lumber yard or home store): Tension Springs
- ☐ (12) Round head #8 wood screws, 1¼ inches long
- ☐ (10 feet) ⅛-inch nylon cord
- ☐ (1) Smaller screw eye, size 212
- ☐ (1) ⁵/₁₆-inch-diameter hardwood dowel, 5 inches long nominal: Trigger
- ☐ String, about 12 inches long
- ☐ Small projectile such as a small rubber ball
- ☐ String, about 6 to 8 inches

TOOLS

- ☐ Small chisel or rotary tool with small straight router bit
- ☐ Drill with ⅛-inch and ¼-inch drill bits and a ¾-inch wood drill bit
- ☐ Hammer
- ☐ Aviation snips or rotary tool with cutoff wheel
- ☐ Wood saw
- ☐ Screwdriver

DIRECTIONS

For this project to work well, we need a large-diameter wooden axle, about 2 inches or more in diameter. But there's a problem—hardware stores don't sell such large-diameter dowels. So, what we'll do is build a large cylinder out of several smaller cylinders.

But how many small cylinders should we use? Topologically speaking, 7 is more or less a magic number when it comes to making a larger cylinder out of equal-sized smaller ones. **Diagram 5.15** shows that a cylinder made out of 4, 5, 6, or 8 identical smaller cylinders is too out-of-round or unstable, but 7 makes a larger cylinder that is just perfect. (In case you're interested, 19 is the next topologically stable number.)

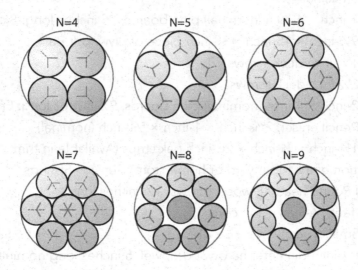

5.14 A large diameter cylinder can be made from smaller diameter cylinders

THE AXLE

5.15 Trigger Grooves

1. Begin by using a router (easier) or chisel (harder) to cut a flat-sided 1-inch-long groove, ¼-inch deep into the same end of 3 adjacent ¾-inch dowels as shown in **diagram 5.15**. These are the Trigger Grooves.

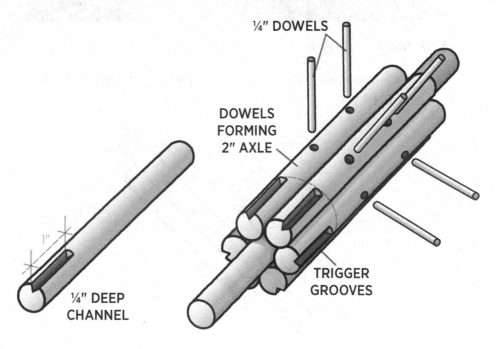

5.16 Axle Assembly

2. Arrange the 7 ¾-inch dowels into a circle as shown in **diagram 5.16**, with the longer dowel in the middle. The longer dowel should extend equidistantly on either side from the pack of shorter dowels. Arrange the trigger grooves so they are oriented outward. Temporarily bind the dowels tightly with masking tape.

3. Drill 6 ¼-inch holes through the axle as shown in **diagram 5.17**. Place glue in the holes and then insert the ¼-inch-diameter, 2-inch-long dowels. Once the glue dries, remove the masking tape.

5.17 Making the Axle

4. Attach the screw eyes to the axle as shown in **diagrams 5.18** and **5.19**.

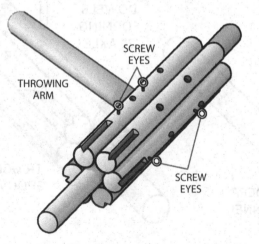

SCREW EYES

THROWING ARM

SCREW EYES

5.18 Attaching Screw Eyes

5. Pound a small nail into the end of the 13½-inch-long, ½-inch-diameter dowel to create the Throwing Arm. Remove the nail head with the rotary tool and cutoff wheel or with aviation snips.

   Drill a ½-inch-diameter hole through the middle of the seven-dowel bundle, on the opposite side of the bundle from the Trigger Grooves, as shown in **diagrams 5.18** and **5.19**, for the Throwing Arm. Insert glue, and then the Throwing Arm, into the hole.

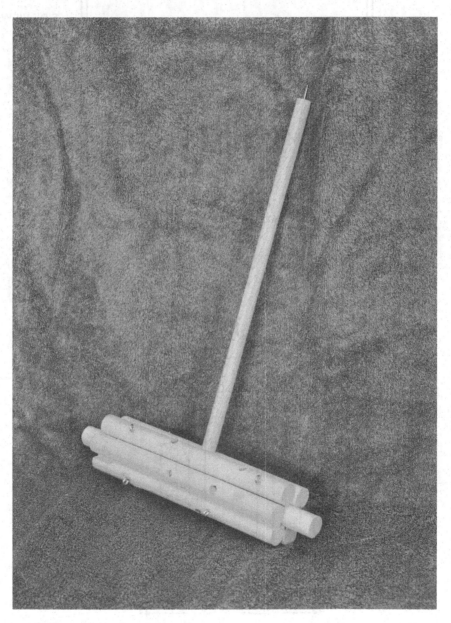

5.19 Completed Axle

THE FRAME

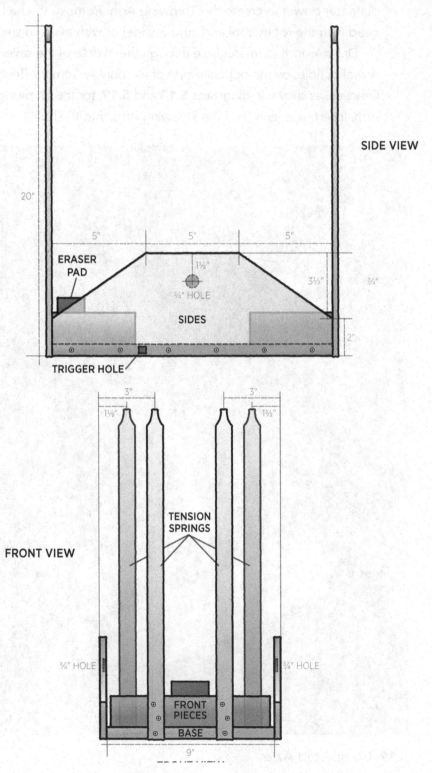

SIDE VIEW

20"

5" 5" 5"

ERASER
PAD

1½"
¾" HOLE

3½"

¾"

SIDES

2"

TRIGGER HOLE

3" 3"

1½" 1½"

TENSION
SPRINGS

FRONT VIEW

¾" HOLE ¾" HOLE

FRONT
PIECES

BASE

9"

5.20 Frame Assembly

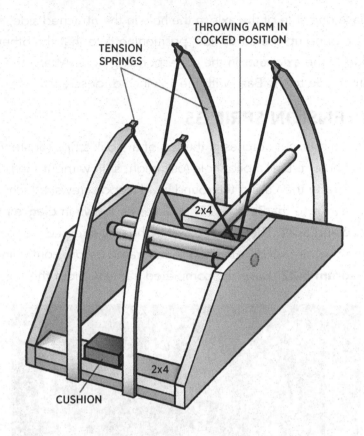

TENSION
SPRINGS

THROWING ARM IN
COCKED POSITION

2x4

2x4

CUSHION

5.21 Frame Complete

6. A sturdy frame is important, so take your time and fasten all pieces together carefully.

   Begin by cutting 15-inch 1 × 6 pine boards into the trapezoid side pieces as shown in **diagrams 5.20** and **5.21**. Then, drill a ¾-inch hole, ⅜-inch deep in the middle of both side pieces, 1½ inches from the top. Do not drill all the way through the wood. These holes act as bearings for the axle. Attach one of the trapezoidal side pieces to the 9½-inch by 15-inch Base using the 1⅝ deck screws.

7. Using the 2½-inch deck screws, attach the 9½-inch 2 × 4 End pieces to the Base as shown in **diagram 5.20**. Glue a rectangular eraser, a sponge, or other resilient surface to one of the end pieces, to cushion the impact of the Throwing Arm when it impacts the frame after the hurl.

8. Place one side of the axle in the hole in the attached side, and then bring up the other side, positioning it so that the other side of the axle rests in the corresponding hole. Attach the other side to the Base with the 1⅝-inch deck screws.

THE TENSION SPRINGS

9. With a saw, cut notches in the top of the oak strips for string attachment. Drill a pilot hole (oak might split without the pilot hole), and then using the round head wood screws, attach the oak strips to the Base and End pieces as shown in **diagrams 5.20** and **5.21**. These pieces will be heavily stressed, so make sure they are solidly attached to the frame before continuing. **Diagram 5.22** shows the completed frame without the Axle.

5.22 Frame without Axle

10. Cut two 36-inch-long pieces of nylon cord. Make a 1-inch-diameter loop in one end of one piece and place it over the notch on one of the Tension Spring ends. Fish the other end of the cord through the two corresponding screw eyes on the axle and then tie another 1-inch loop and affix the cord to the top of the Tension Spring on the same side of the axle. Do the same for the Tension Springs on the opposite with the other piece of cord. See **diagram 5.23**. Take time to adjust the length of each cord so there is little slack in the cords prior to rotating (or "cocking") the throwing arm.

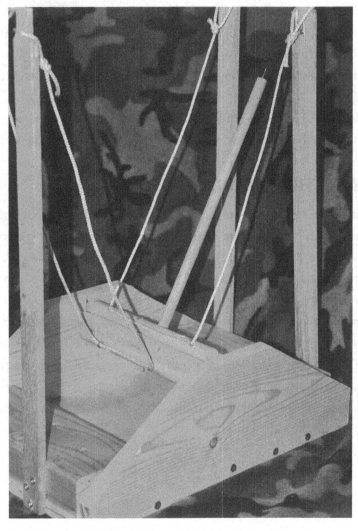

5.23 Tension Spring Cords

11. Adjust the tension in the cords by untying and retying so they are nearly equal when the Axle supporting the Throwing Arm is rotated until it makes contact with one of the End pieces. You can gauge the tension by plucking the string and listening. When all cords have the same pitch when the Throwing Arm is down in the cocked position, then the tension in the cords is equal.

THE TRIGGER

The Trigger is a small dowel that jams between the Trigger hole and the grooves drilled in step 1. In Leonardo's machine, he used a toothed wheel and cam to allow him to hold the Throwing Arm in place before firing. But that's complicated to make. Luckily, using a trigger stick works nearly as well and is far simpler to construct.

12. Tie the 12-inch-long string to the middle of the 5⁄16-inch-diameter hardwood dowel, which will act as the Trigger dowel.

FIRING THE CATAPULT

13. Cock the device by grasping the Throwing Arm and rotating it until it just contacts the End piece without the eraser glued to it. As you turn the arm, the Axle rotates, wrapping the cords around it and pulling the oak Tension Springs. Keep a firm grip on the arm because you don't want the arm to whip around and hurt you. When the machine is cocked, place one end of the Trigger against one of the front pieces and the other end in one of the grooves cut into the axle assembly. See **diagram 5.24**. Adjust the trigger stick so that it is easy to pull out with the string, but not so precarious that the catapult could fire accidentally.

5.24 Trigger

14. Make a projectile by taking the thing you want to throw (a small rubber ball or something similar) and attaching a string about 6 to 8 inches long with a loop on the end.

15. Place the loop around the nail on the end of the Throwing Arm and arrange the projectile so it sits on the End piece. Now, aim the catapult. Make sure there is nothing in front (or in back, more on this later) that you don't want hit with the projectile. This includes breakable objects, windows, animals, people, and so forth.

16. Pull smartly on the string to remove the Trigger dowel and fire the catapult. You can adjust the height and range of the catapult by shortening or lengthening the string. Make it too long and the projectile will bounce on the ground just in front of the catapult. Make it too short and the projectile could go backward!

17. Now, it's time to readjust the tension in the Tension Springs by making the cords longer or shorter. Retie the cords around the oak Tension Springs as necessary. You will want the tension in each spring to be more or less equal when the machine is ready to fire. Do this by cocking the machine back, securely inserting the Trigger dowel, and then plucking the dowels and listening to the sound each cord makes. They should all have the same frequency. If one is higher or lower pitched than the others, retie the cords.

Don't overstress the springs or they could crack. As always, use this catapult at your own risk.

5.25 Completed da Vinci Catapult

5.26 Catapult Cocked and Ready to Fire

6

PETER THE HERMIT AND HIS CRUSADERS

After reading about Genghis, Attila, and Tamerlane, you might jump to the conclusion that medieval invaders always traveled westward from the steppes of Central Asia to ravage and pillage the ostensibly peaceful and law-abiding peoples of Europe. This is untrue on multiple counts. First of all, the inhabitants of Medieval Europe, looked upon as a whole, were hardly peaceful and righteous. Second, they were as aggressive toward their eastern neighbors as the earlier Huns and later Mongols were to them.

The history of the Crusades proves the point. In the late 11th century, Pope Urban II preached, cajoled, and commanded from his pulpit, calling for a volunteer army of Christians to take back Jerusalem from the Seljuk Turks. It was wrong, he said, for Jerusalem and other places holy to Christians to be under the power of non-Christians. God, he said, did not want this to be. His words resounded in the ears of Europeans, rich and poor alike. But not necessarily for religious reasons.

For the rich, Urban's request seemed like a good excuse for empire building, or, at least, it was a chance to gain power and influence. Besides the chance to fight for their religion (fighting for any reason was something medieval knights loved to do), the rich lands of the Levant were a prize worth the bother.

But for the poor, the prize was even greater. Urban's call was a chance to escape the grinding tedium of their nearly slave-like existence. It was everything poor, bored, starving peons could want: a pope- and God-approved excuse to leave behind their miserable lives toiling in the fields for starvation wages, take up the sword, and make a fortune by plundering and pillaging in the land of infidels.

In 1095, Pope Urban II spoke to a vast multitude assembled in the market place of Clermont, France. They were excited to hear his speech, for it was a clear and unambiguous command from God, voiced by his earthly representative, to take back the lands that were so rightly theirs. Urban ascended his lofty pulpit and addressed his impatient audience. He spoke well and eloquently. Time after time, he was interrupted by the clamorous shouts of all those assembled: Deus vult! Deus vult! (God wills it! God wills it!)

"It is indeed the will of God," shouted the pope. "Let this memorable word . . . be forever adopted as your cry of battle. . . . His cross is the symbol of your salvation; wear it; a red, a bloody cross, as an external mark on your breasts or shoulders, as a pledge of your sacred and irrevocable engagement."

"Deus vult" became the battle cry of these men, soon termed Crusaders. Beginning in 1096, successive waves of Crusaders took up the crusade against their hated, but unmet, enemies. This included the Turks, of course, but also included the Jews, Eastern Europeans, the Byzantines, or just about anybody who had something they wanted.

The very first wave to head east was known as the Peasants' Crusade. They were a ragtag bunch of irregular soldiers, men freed from their villages and farms by Urban's call to march east in hope of finding rich plunder and a guaranteed ticket to heaven should they die during the fight. Many were dirt-poor farmhands; some were half-starved street beggars. Others were thieves and murderers, given a reprieve from the gallows by their promise to kill as many Turks as possible during their trip east. All in all, it was a very motley and vulgar assemblage at best.

Their leaders were a mixed bunch, ranging from a French knight called Walter the Penniless, to a man claiming to be the long-dead Charles the Great who had been raised from the dead to lead the expedition. Of all the leaders, perhaps most important was a half-crazed monk called Peter the Hermit.

6.1 Peter the Hermit   *J. G. Edgar*, Danes, Saxons, and Normans; or, Stories of our Ancestors *(1863)*

Although some later historians take a differing view, chroniclers of the time wrote that Peter the Hermit was the key organizer of the Peasants' Crusade. For example, Roger of Wendover and Matthew Paris, both well-known but usually ill-informed contemporary writers, tell of Peter's pre-Crusade visit to Jerusalem during which Jesus appeared to the Hermit in the Church of the Holy Sepulchre and instructed him to lead the crusade. Although by modern standards he seemed positively delusional, Peter was still given an audience with Pope Urban II.

A short time later, German crusaders from the Rhine Valley went one better in terms of finding a nontraditional leader: they followed a goose believed to be enchanted by God as their guide east. The goose, however, soon died, and so they joined up with Emicho of Leisingen, who loudly proclaimed to anyone that would listen that a cross "miraculously" appeared on his chest, proving he was God's selection to lead them.

Emicho and his followers expanded the scope of the mission given them by Pope Urban and decided that they should go beyond the Turks and eliminate all non-Christians in their path. The crusaders proceeded to massacre the Jews in places such as Lorraine, Mainz, Cologne, and Worms. Thousands of defenseless men, women, and children were killed.

As a whole, the vast army of the Peasants' Crusade turned out to be anything but holy warriors. When Peter the Hermit's army entered Yugoslavia, 4,000 Christian residents of the city of Semlin, now a suburb of Belgrade, were massacred before the Crusaders moved on to burn Belgrade. Crazed with religious zeal mixed with bloodlust, the army killed thousands of innocent people as they marched east, well before they saw their first Turk. They brought their mayhem upon the poor inhabitants of Hungary and Bulgaria, stealing, pillaging, and killing as they went. But the local Hungarians and Bulgarians gave as good as they got, and killed thousands of Crusaders. Badly whipped by the better prepared peoples of Eastern Europe, the remaining Crusaders (and there were still many of them) eventually straggled into Constantinople, the capital of the Byzantine Empire.

Their reputation preceded the Peasants' Crusaders so the Byzantines were not particularly pleased to see them arrive. But Byzantines were powerful enough to contain them in temporary camps outside the city, and clever enough to know what to do with them.

In the autumn of 1096, Alexius I Comnenus, the Byzantine Emperor, addressed Peter and his horde of smelly, undisciplined, berserker peasants camping on his doorstep. He told them it was time to move east to engage and conquer their real enemy, the Turks. This, at long last, was the order they had been waiting for, and they immediately obeyed it. Raised to new heights of anti-infidel fervor, the zealots crossed the Bosporus and swarmed into the Turkish kingdom of Roum.

As the Byzantines surely knew, this group, despite their size, was no match for the regulars of the Turkish army. At the castle of Exorogorgon, their opponents cut the Crusaders to pieces. The Turkish infantry rained clouds of arrows down upon them, and mounted cavalry speared the Peasants' Crusaders pretty much at will. The few not killed were sold into slavery or used for archery practice. The sly Alexius was very likely not at all displeased with the outcome.

DEFENSIVE EARTHWORKS

From Neolithic times onward, home defenders have shaped and mounded the landscape around their homes to make them more secure. Even in the earliest societies—the Druids in England, the mound builders of pre-Columbian America, and the Mesopotamians who lived in Jericho, for example—people have built earthworks for protection.

There are many examples of early defensive earthworks. Silsbury Hill, a 130-foot-high mound on a plain near Stonehenge, was built by ancient Britons around 5,000 years ago and is assumed by many to have been built to support both warlike and religious affairs.

Similarly, ditches have a very long history. In Austria, archaeologists recently uncovered a Neolithic settlement covering a full 50 acres. This place, a few miles north of modern Vienna, is known today as Grossrusbach-Weinsteig. The early human settlement was protected by a 12-foot-wide defensive ditch. Radiocarbon dating estimates that the ditch was constructed around 4500 BCE.

The most effective protection method in past times was the water-filled trench, or moat. Even today, despite the great strides made in home security and self-defense technologies, moats still provide a level of protection unrivaled by any other defensive earthworks. While building a moat around your castle is neither inexpensive nor simple, that plain fact is that, once built, a moat assures a level of protection against invaders and interlopers like nothing else.

The 12th through 15th centuries were the salad days of moat building. Hundreds of moats were installed in England, continental Europe, and the Far East. Few moated buildings were actually attacked and fewer were taken, attesting to the psychological and physical advantages a moat provides.

6.2 Castle and Moat   *Copyright Simon Ledingham*

Consider the siege of Rhodes castle in 1480. An Ottoman army attacked a garrison of about 3,500 Knights Hospitaller inside the fortress. Although it is written that more than 70,000 Ottoman soldiers besieged the castle, the design of the fort, with its thick walls, colossal towers, and above all, its gigantic dry moat, was impervious to repeated attack.

Crusader-Proof Moat

Today of course, moat construction is no longer common or easy. A large, well-built moat could cost $50 or more per square foot, although costs vary widely depending on factors such as the amount of excavation required, the soil conditions, and the availability of water.

Before you commit to building a moat around your home, consider the following:

- You'll need a dependable water source to fill and maintain the moat. A 12-foot-wide moat, 200 yards long and 6 feet deep, will require an initial fill of nearly 500,000 gallons. Although soil composition varies from site to site, most moats should be lined with layers of clay to stop water seepage.

Diverting natural sources of water such as streams and springs is the most economical method of obtaining water, but be sure to consider the attendant legal and environmental issues prior to building the moat. If possible, connect your moat to a source of running water such as a stream via movable sluice gates. Flowing water will reduce the problems (e.g., odor, insects, and pond scum) associated with stagnation.

If no natural source of water is available, wells and municipal water systems are alternatives but can be very expensive. Talk with an attorney experienced in water rights before investing in any moat or moat-related equipment.

- Consider your climate. In cold weather areas, the moat could freeze over in winter, providing easy access to your property for people or even vehicles, negating the moat's most important security features.
- Contact your local zoning commission before beginning construction. In many cases, a variance or special permission of your neighbors may be required. A building permit is likely required.
- Few modern building contractors have experience building moats. Carefully interview prospective contractors and make sure they have completed projects that have included large water-holding structures in the past. It's generally best not to have builders learn new techniques on your dime.
- Moats are high-maintenance items. In hot months, stagnant water can lead to high insect populations, deteriorating masonry, offensive odors, and unsightly trash buildup unless the moat owner commits to a regular program of moat maintenance. As evidenced at the fortress of Rhodes and Nagoya Castle in Japan, it is possible to forgo water and build what is termed a "dry moat." However, the physical and psychological protections afforded by a water-filled barrier make wet moats more effective.
- If possible, consider locating your home on a small, isolated island instead. Doing so, a home owner can accrue all of the security advantages of moat ownership without the high initial outlay and ongoing maintenance costs.

These drawings will provide a starting point for building a moat around your home. Moats are most effective when paired with a high masonry or stone wall, called a curtain wall, just inside the moat. The curtain wall should be at least 8 feet high and 4 feet thick at the top and extend, save for the gatehouses, continually in an unbroken line around your property. These dimensions will provide adequate space for a maneuvering area, known as a rampart, atop the wall for defensive action should the situation arise. The ramparts should be protected by battlements, which are walls with regular openings used to provide cover while allowing offensive fire. Curtain walls may be made from any locally available stone provided the stone is free from pits, fissures, and other water entry points. Granite, limestone, and flint work particularly well.

At least two entry points or gates through the curtain wall are required. One is the main entrance and the other is for emergency egress should the main entrance become unusable. Gate planning is especially important when planning a moated home space. Place access stairways to battlements and ramparts at frequent intervals. Stairways should be made large and wide enough to allow material and ammunition to be carried conveniently to the battlements.

When constructing the moat walls, there are many options to consider. It is important that the contractor use iron rebar inside the concrete as reinforcement as the pressure applied by the water to the sides and bottom of the moat is substantial.

Vintage castles often used several layers of clay hand-applied over the excavation area. However, modern construction methods are definitely superior. "Shotcrete" or "gunite" construction is the most cost-effective method of lining a moat. Both of these terms refer to the same concept—a pneumatic method of concrete application that uses high-pressure air to spray a mix of marble dust, mortar, and water through a hose onto the surface of the moat.

A moat construction crew digs a hole, installs the plumbing necessary to fill it with water, and then assembles a framework grid of ⅜-inch steel reinforcing rods (rebar) covering the bed of the moat. The rebar are laid out on 10-inch centers and fixed in place with wire.

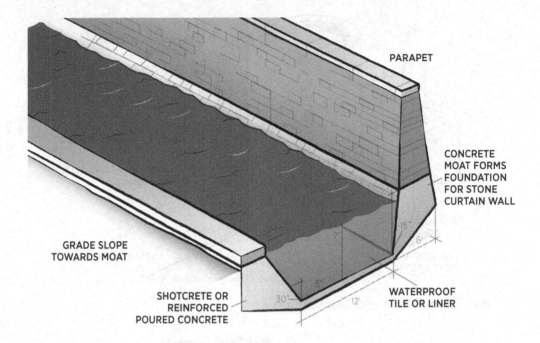

PARAPET

CONCRETE
MOAT FORMS
FOUNDATION
FOR STONE
CURTAIN WALL

GRADE SLOPE
TOWARDS MOAT

SHOTCRETE OR
REINFORCED
POURED CONCRETE

WATERPROOF
TILE OR LINER

6.3 Moat Cutaway

The crew sprays a heavy coating of shotcrete mix around the rebar. The crew trowels the shotcrete and lets it cure for a week. Then the crew returns to apply a second finish coating. If desired, stones or pebbles can be embedded in the shotcrete to give it a better-looking finish. A moat built in this fashion will be long lasting and impervious to most mining or escalade attacks.

Alternatively, a moat may be constructed from poured concrete, but this type is generally more expensive to build. Instead of spraying concrete material around a rebar framework, concrete is poured into wooden forms and allowed to cure.

Egress and ingress to the home through the moat and curtain walls is made through specially designed gates and bridges. There are many important considerations when designing curtain wall fenestration, including the drawbridge design, bridge operators, and gate house construction techniques. **Diagram 6.4** illustrates one type of drawbridge. It is important that the drawbridge be designed such that it can be operated manually (e.g., via a hand winch or block and tackle) should electricity become unavailable. Be certain to discuss these items carefully with your moat builder prior to groundbreaking.

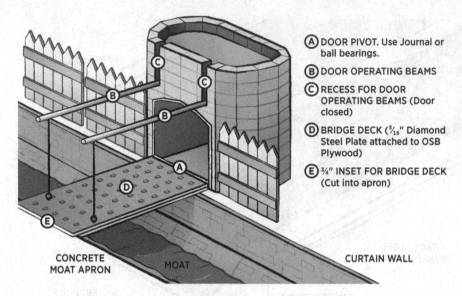

A DOOR PIVOT. Use Journal or ball bearings.

B DOOR OPERATING BEAMS

C RECESS FOR DOOR OPERATING BEAMS (Door closed)

D BRIDGE DECK ($\frac{3}{16}$" Diamond Steel Plate attached to OSB Plywood)

E $\frac{3}{4}$" INSET FOR BRIDGE DECK (Cut into apron)

CONCRETE MOAT APRON

MOAT

CURTAIN WALL

6.4 Gate House and Drawbridge

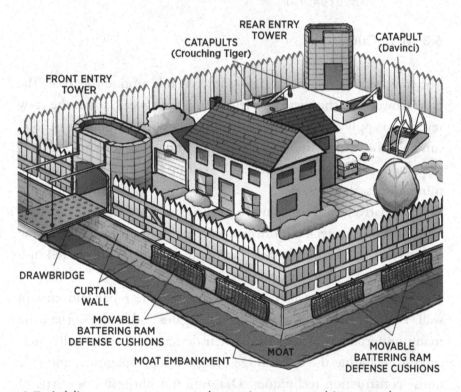

REAR ENTRY TOWER

CATAPULTS (Crouching Tiger)

CATAPULT (Davinci)

FRONT ENTRY TOWER

DRAWBRIDGE

CURTAIN WALL

MOVABLE BATTERING RAM DEFENSE CUSHIONS

MOAT EMBANKMENT

MOAT

MOVABLE BATTERING RAM DEFENSE CUSHIONS

6.5 Adding a moat, towers, battering ram cushions, and catapults will make even the most determined Crusader or Tatar think twice about messing with you.

RAGNAR LODBROK AND THE VIKINGS

One of the most famous Viking raiders was the Danish tough guy Ragnar Lodbrok. He was also known as Ragnar Hairy-Trousers because he was said to wear pants made of rough wool. According to the Norse sagas, among the many of Ragnar's sons were the gruesome-sounding Sigurd Worm-in-Eye and the slightly less gruesome-sounding Ivar the Boneless. It is written that poor Ivar was born without bones in his legs and had to be carried to and from battles. Despite his disability, he was evidently quite an able fighter. According to the sagas, he would order his assistants to raise him from the large shield on which he was usually carried and more or less hurl him into the fray. In one episode he landed upon and killed a magic cow that protected Swedish King Östen. (Perhaps this cow was related to the enchanted goose of Crusader fame.) In any event, while a legless Viking chief being catapulted into battle sounds more than a bit cartoonish, the sagas clearly proclaim Ivar a successful and able Viking warrior.

Starting in roughly 800 CE and continuing to about 850 CE, Ragnar fought in 51 battles, compiling an enviable record of 50–1. One of his early raids consisted of leading 5,000 men in Viking longboats down the River Seine to Paris, where he coerced Charles the Bald into paying a tribute of 7,000 pounds of silver to make the Vikings go away. Finding the protection racket an easy way to make a lot of money, he and his Viking raiders sailed up and down the English and French coasts threatening, intimidating, demanding, and getting a lot of silver and gold.

7.1 Viking Siege of Paris   Der Spiegel Geschichte

It was a good life and a pretty easy one, since most of the time Viking victims chose to simply pay up before things actually got physical. But unfortunately for Ragnar, such activities ran afoul of the Danish King Horik, who was angry with Ragnar—not so much for his lawlessness as for his unwillingness to share the loot. He put his kingly foot down and tried to arrest him. But Ragnar and his sons escaped from Scandinavia and turned their attention to the easy pickings on the coasts of Ireland and England.

Again, for several years, the raiders lived well. But Ragnar pressed his luck once too often. On the 51st raid, he met his match. Aella II, the

King of Northumbria, captured Ragnar. According to different accounts, it happened either in battle or when his dragon-shaped longboat capsized in a storm. In any event, Ragnar was brought before Aella, who, it is written, ordered him cast into a pit filled with poisonous snakes. Thus ended the life of one of the most colorful Viking raiders.

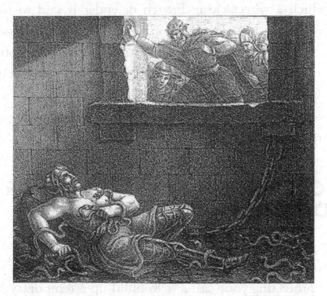

7.2 Ragnar   *Hugo Hamilton,* Teckningar ur
                                        Skandinaviens Äldre Historia *(1830)*

But Ragnar's demise didn't mean that the Viking raiders stopped their attacks. In 885, under command of Siegfried, the Vikings again sailed down the Seine toward Paris. At the time, Paris was a small town mostly situated on a single island in the middle of the Seine. But it was strategically located and the city defenders, led by Count Odo and Bishop Gozlin, blocked the river and wouldn't allow the Nordic armada to continue. In the intervening years since Ragnar's last visit, the Parisians had bolstered the city's defenses by building forts to protect the bridges that linked the east and west banks of the city.

Siegfried's Vikings were a force of 30,000 men carried in 700 ships, although estimates to the actual size vary, and those numbers seem a bit large to swallow. Stopping outside the city walls, Siegfried demanded money, and lots of it, in return for leaving Paris alone. The French, according to the ancient chronicles, had but 200 men inside their walls; hardly enough, it seemed, to even slow down the Danish advance.

But through luck and determinations, Count Odo and his plucky French defenders held off the attackers for nearly a year, withstanding nearly daily missile attacks from mangonels and other siege weapons. In response, the French built their own catapults, one of which fired a heavy spear that is said to have transfixed seven Vikings like a shish kebab, which a wisecracking French defender is said to have then ordered to the kitchen. The French maintained a stiff defense, and in the end the Vikings gave up and simply dragged their longboats overland to their next target in Burgundy, bypassing Paris entirely.

For Odo, knowing a thing or two about catapults and siege defense worked out quite well. Called a hero for his bravery and skill, he was crowned king of the Franks in 888 CE.

THE CHEVAL-DE-FRISE AND OTHER FIELD EXPEDIENT DEFENSES

Imagine you hear of a rapidly approaching Viking army. Assuming that Vikings outnumber you in people and hand weapons, your best hope for protecting your castle is to build up a temporary but solid defensive line and hunker down until help arrives. Such measures are known in military jargon as "field expedient defenses."

The famous Civil War Union general William Tecumseh Sherman wrote that temporary defenses like these "enable a small force to hold off a superior one *for a time*, and time is the most valuable element in all wars." If you find yourself in such a position, you'll be in far better shape if you have a basic knowledge of the art of erecting defenses built quickly from materials that don't require much preparation, such as earth, brushwood, and light timber.

Since Hun, Mongol, and Viking threats arise quickly, often you won't have the time and resources to build permanent fortifications such as moats or palisade walls. But it is possible to erect temporary obstacles to hinder or even thwart dangerous raiders. As Sherman noted, obstacles themselves merely slow an attacker's advance. They must be used in conjunction with offensive weapons.

The first item of business in constructing a solid defense is to reduce the number of routes from which a Viking or other raider can approach your castle. This means all trees, hedges, and buildings around your castle that could provide cover for raiders should be removed as soon as a real and credible threat to your castle is detected. Leave no stump more than a foot high; leave no hiding place or defensive redoubt for the cunning Hun to crouch behind. True, your neighbors may object to the removal of their trees and buildings, but in the case of a Hun attack, it's better to lose trees than lives, so extreme measures are justified.

Next, assuming your basic defensive position is your castle, remove curtains and other flammable materials from the window area facing the expected direction of attack. Fire, caused by incendiary projectiles flung from catapults and fire arrows, is an important concern. Containers of water should be placed in each room. Place sandbags in windows to narrow the opening, but leave a firing port.

Once the approaches to your castle are cleared and the house prepared, it's time to construct and install your field expedient defenses. The best expedient defenses are large trenches and ditches. However, the design and digging of such excavations are complicated and beyond the scope of this book. But there are several types of quickly erected defensive structures that are buildable with a minimum of time and money. These include the trous-de-loup (pits), abatis (tree limb obstacles), and chevaux-de-frise (wooden barricades).

Why are most of these terms French? In the period in which these barriers were invented, France had the most well-respected army in the world. Many of their tactics and strategies came directly from Napoleon and his generals.

TROUS-DE-LOUP

Trous-de-loup roughly translates to "wolf-holes." They are military pits made in two ways, either deep or shallow. Deep pits are typically 6 feet or more deep, and 6 feet wide at the top, narrowing to a single foot or so diameter at the bottom. Such a pit provides quite an obstacle to any attacker. Several offset rows of deep trous-de-loup, equipped with a set of sharpened stakes at the bottom, provide a formidable defensive obstacle.

In a similar vein, hastily dug shallow pits can effectively slow both cavalry and infantry charges. (Note well: Unless you're in imminent danger, you should not dig trous-de-loup. These are effective but very dangerous!)

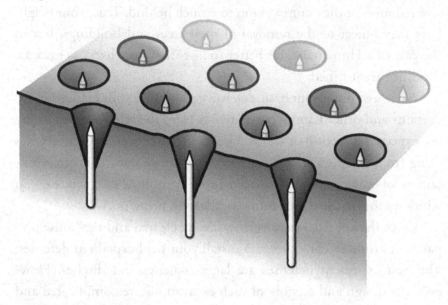

7.3 Trous-de-loup

ABATIS

An abatis is a quickly constructed defensive obstacle made from felled trees. The idea is to make a thick, difficult-to-traverse matrix of branches that will slow or prevent both Hun and Tatar horses and Crusader and Macedonian infantry attack.

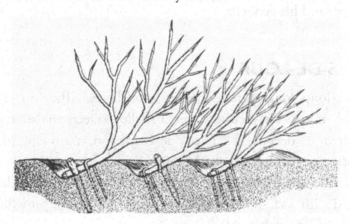

7.4 Abatis

US Army TM E 30-480 Handbook On Japanese Military Forces (1944)

Making an abatis is actually quite simple. Chop down several trees and place them in the likely approach to your castle. Orient the trees such that their branches interlace, with the tips of the larger branches pointing toward the likely direction of Hun attack. If time permits, remove the smaller branches from the felled trees, sharpen the ends of the thicker branches pointing in the direction from which the attack is expected, and anchor the main trunk to the ground using spikes to make the structure more difficult to move or push away. Such a field expedient defense can slow a ground attack considerably.

CHEVAUX-DE-FRISE

Frederick the Great, the famous 18th-century king of Prussia, was a keen student of military tactics. He wrote extensively on this subject, and his writings are still considered foundations of modern military strategy. In one of his treatises, he describes the chevaux-de-frise, which were pieces of timber or logs about a half foot in diameter and 10 feet in length. Sunk into the log are closely spaced, perpendicular rows of sharpened wooden stakes, 10 feet long, set so they extend equally from both sides of the main timber.

Frederick writes that chevaux-de-frise were first used at the Battle of Groningen in 1672 between the French and the Dutch. At that time, the people of the Dutch state of Friesland found themselves in a precarious position—their country was being invaded by an army of French horsemen and they had no cavalry of their own. So, some brilliant but now unknown officer among them concocted one of the most well known of all field expedient defenses on the spot. Dutch lines of bristling wooden spikes made it impossible for the horses of the invaders to make any headway against them. With an ironic sense of humor, the French soldiers dubbed the device "the horse of Friesland," or cheval de Friesland. This was soon corrupted to cheval-de-frise, or, in the plural, chevaux-de-frise.

Military engineers took note of the device's effectiveness and it soon became a standard method of protecting military lines. For example, at the Battle of Petersburg during the American Civil War, both Union and Confederate soldiers built a long line of Friesian Horses to reinforce their defensive lines. The men of the 50th New York Engineer Regiment worked night and day, often under heavy fire, to construct a wall of such obstacles to keep a large rebel force from attacking. Photos of the sturdy, sharpened spikes show it to be a fearsome-looking obstacle, and one that neither infantry nor cavalry soldiers could easily overcome.

It's time once more to consider a counterfactual scenario. Had the techniques of building the cheval-de-frise, trous-de-loup, and other field expedient defenses been developed at an earlier point in history, would the world look different? Specifically, what difference would such constructions have made against Hun horse-mounted attacks in Gaul and Italy or against Mongol riders in Russia and China?

Quite possibly, easily constructed but effective defenses such as the ones described in this chapter would have forced a major change in the tactics of the invaders. They would have been forced to leave their beloved horses and slog into battle on foot where they were far less effective warriors. Had this happened, perhaps the Slavic kingdoms of the 13th and 14th centuries would have grown and prospered rather than withered as they did when defeated by the Mongols. To the east, the various Chinese clans might have remained independent had they not been subjugated by the Mongol horsemen. China, instead of the single country we know today, may have become several smaller states. Quite possibly, the map of the world would look far different.

7.5 Cheval-de-frise at Petersburg

Library of Congress LC-DIG-cwpb-02597

The Cheval-de-frise

The great thing about a model cheval-de-frise is that it is easily built with hand tools and can be scaled up to life-size using the same construction principles if desired. As described below, this model makes an interesting pencil holder/desk accessory and history-related DIY project, and one sure to get attention from visitors!

MATERIALS

☐ (1) square wooden dowel, 1-inch × 1-inch × 10-inches long

☐ (18) round (not hexagonal) wooden pencils, sharpened (Mirado Black Warrior pencils are sold in many US office supply stores. If you can't find any at your local store, use this Internet search term: "round pencils.")

TOOLS

☐ Ruler

☐ Pencil

☐ Drill press (a hand drill will work, but it takes more care to make a perfectly straight hole) with a $5/16$-inch drill bit

DIRECTIONS

1. Using your ruler and pencil, carefully lay out the hole drilling pattern shown below on the dowel. Make a mark for each hole, laid out on the centerline of the long faces, with each hole 1 inch apart from the next parallel hole, and ½ inch apart from the adjoining perpendicular hole, as shown in the drilling diagram.

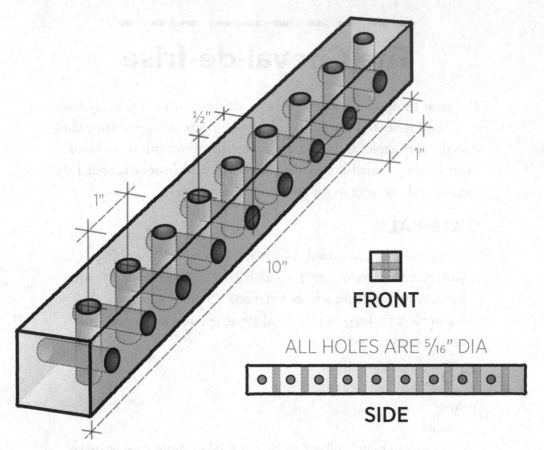

½"

1"

1"

10"

FRONT

ALL HOLES ARE ⁵⁄₁₆" DIA

SIDE

7.6 Cheval-de-frise Drilling Diagram

2. Drill 9 holes in each face of the dowel. Keep the drill bit perpendicular to the wood face and drill ⁵⁄₁₆-inch diameter holes through the square dowel, according to the drilling pattern laid out in step 1.

3. Insert the sharpened pencils into the holes in the square dowel. You can easily adjust the position of the pencils (the spikes) up and down in their holes, and in so doing, change the angle at which they point. A 50-degree angle works well, but you can adjust the angle depending on what barbarian horde is attacking you. For example, if you are defending against, say, a Macedonian phalanx, then arrange the spikes so they point at a more acute angle from the ground. If defending against a human soldier versus Mongol pony, then tip the cheval-de-frise so the spikes point higher.

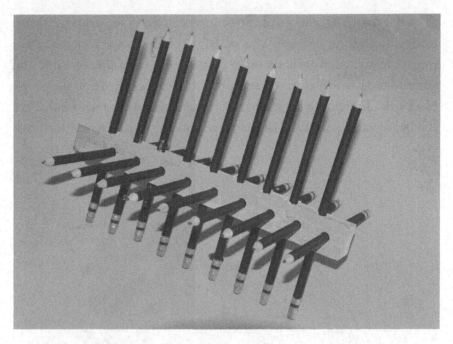

7.7 Cheval-de-frise

BELL TOWERS, CIVIC TOWERS, TOWER HOUSES, AND WATCHTOWERS

Observation towers, in all their forms—bell towers, civic towers, tower houses, and the watchtowers of city walls—were common throughout the ancient and medieval world. Many were so solidly built that even today in the lands where the invading armies of Alexander, Attila, and Genghis once ran amok, the civic watchtowers still stand and beckon visitors.

A watchtower can provide an extra measure of security for your castle. When belligerent hordes start massing themselves for an attack, every second counts. The earlier preparation for defense begins, the better are your chances for coming through the unpleasantness in good shape. So, consider erecting a watchtower or observation tower on your property. The elevated platform provides a spot from which you can see a long way off, and also provides "high ground" for the use of crossbows and, if you have them, gunpowder weapons.

Diagram 7.8 shows a watchtower constructed in the field by Russian troops during the Crimean War of 1853. Simple towers such as these were constructed at frequent intervals along the front line of battle and enabled the Russian observers at the top to keep a close eye on the course of a battle. To signal to the troops below them, the tower-men would light torches of various sizes and shapes corresponding to different situations.

7.8 Crimean Watchtower   *"Military Watch-Towers in the Crimea,"* The *Illustrated Magazine of Art* Vol. 4, No. 23 (1854)

With a little thought and preplanning, we can construct watchtowers as well, if not as tall as the Tsarist army's. A well-placed, well-made tower would provide a great advantage in staving off a barbarian attack.

How tall should your watchtower be? That depends on many factors, and on your situation. For the scientifically minded, you can easily figure out how much farther you can see by standing atop a watchtower of a given height, assuming a flat or nearly flat horizon. Just plug your tower height (plus the height of your eyeballs above the viewing platform) into this equation:

$d \approx 1.22 \times \sqrt{h}$ where d is the distance you can see in miles, and h is your height in feet

For example, if you stand about 6 feet tall, you should be able to see:

$$1.22 \times \sqrt{6} \approx 3 \text{ miles (over flat land or water)}$$

If you stand on a 10-foot-high watchtower, then you increase your field of view to:

$$1.22 \times \sqrt{16} \approx 5 \text{ miles}$$

And if Viking dragon boats appear, seeing them coming from two extra miles may make all the difference.

Choosing the site of your watchtower is an important consideration. Obviously, it must provide as full a field of view as possible, yet it must be defendable or at least lend itself to an easy exit to a safe location since it is likely the first target for opposing forces to aim upon.

The Hourglass Tower is an easy-to-build watchtower. It provides a fairly sturdy (depending on the care with which it is erected) platform for scanning the horizon for threats. The tower goes up fairly quickly, and while it can be erected in a few hours by a single builder, the task is far easier if a helper or two can be recruited.

The watchtower design and construction techniques that follow are, similar to those for the Crimean War watchtower illustration, field expedient. The tower is inexpensive, requires little in terms of site preparation, goes up quickly, and comes down even faster. Because of the way it is constructed, it's unlikely that your municipality would require a construction permit, but it can't hurt to check. Please note that while the materials are quite strong, the tower has not been tested for all conditions. Like all the projects in this book, if you build and use it, you do so at your own risk.

Nowadays, building such constructions of wood and rope is pretty much the purview of Boy Scout troops and back country campers, although at one time, such skills were de rigueur for military engineers. In fact, the instructions that follow are modified from information in the *US Army Survival Manual.*

The Hourglass Watchtower

This design is based on the lashing techniques learned by nearly every Boy Scout in order to earn his first-class scout rank. Lashing involves connecting together poles (called "spars" by those in the know) with rope, using specially devised knots that make the connections fast and rigid. There are several advantages to building structures by lashing, including the facts that the materials are quite inexpensive, few tools are necessary, and the materials can be easily reused to make something else when the need for the current project has passed.

The hourglass tower, so called because the basic superstructure resembles an hourglass, goes up quickly and is quite stable if care is taken to make it as symmetrical and balanced as possible and the lashes are made correctly.

7.9

Construction notes:

- The lower tripod in the hourglass must have a wider base than the upper tripod to provide a stable platform less likely to overturn.
- The structure should be securely staked to prevent tipping, using the staking technique described in the project.
- Manila rope is very hard on the hands. Wear gloves when pulling tight to avoid blisters and skin irritation.
- Wrap a small piece of duct tape around the ends of each rope to prevent fraying.
- Ropes will stretch and spars will settle over time, loosening the lashed connections. Inspect your tower frequently and retie the lashes as required.

Safety Considerations

- Use great caution when erecting your tower and even more when using it. Falling off the platform could be very dangerous. Also, since this is a home-built, field expedient construction, the quality of the materials and the techniques used to lash together spars and tie knots could be less than perfect. If any part of the structure fails (a spar snaps or a rope breaks), the tower could fall, causing injury or worse to those on the tower as well as those below it. Carefully inspect all materials used for spars for defects and discard any piece that does not meet your standard of quality. I built this structure using 2-inch-diameter bamboo and it held me easily. I can't provide an estimate for how much your tower will hold because of the variability in equipment and supplies, so carefully test your tower before using it.
- Spars for your tower may be made from many different types of wood. (I have never made spars made from metal or plastic, although I imagine that it could be possible.) Typically spars are thin, strong lengths of wood, around 8 feet long and 2 to 3 inches in diameter. Green wood

is less likely to snap than dry wood. I used bamboo with good results. Bamboo poles are very light, strong, and fairly inexpensive. The most common types of bamboo used in lashing projects are Tonkin and Moso varieties, and they stand up well to the weather. To find mail order sources of bamboo, type "8 foot bamboo poles" into your Internet search.

■ Use ⅜-inch-diameter manila rope or a synthetic equivalent made from polypropylene. Manila rope is made from the stems of the abaca plant. It is strong, durable, organic, and cheap. One of the things that make it excellent for lashing is that is very rough textured. While that makes it attach securely to the spars, it is very hard on the hands, so leather gloves should be worn when pulling on the rope to make the lashes. (You'll likely need to remove your gloves to tie the some of the required knots, however.)

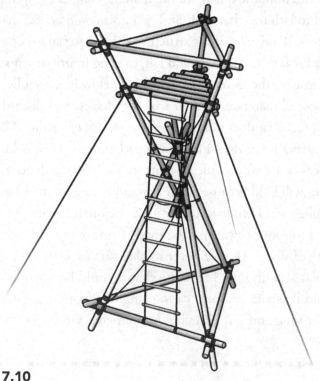

7.10

MATERIALS

- ☐ 400 feet of ⅜-inch-diameter manila rope
- ☐ Duct tape
- ☐ (6) 2-inch-diameter bamboo poles 8 feet long: Spars
- ☐ (6) 2-inch-diameter bamboo poles 8 feet long: Poles
- ☐ (3) 2-inch-diameter bamboo poles 4 feet long: Platform Rails
- ☐ (6) 2-inch-diameter bamboo poles 2½ feet long: Platform
- ☐ (7) 2-inch-diameter bamboo poles 1 foot long: Steps
- ☐ (9) Approximately ¾-inch-diameter wooden stakes, 2 feet long
- ☐ (6) Approximately ¾-inch-diameter windlass poles, 18 inches long

TOOLS

- ☐ Leather gloves
- ☐ Large knife or scissors
- ☐ Tape measure
- ☐ Saw
- ☐ Hammer for pounding stakes

DIRECTIONS

General Notes on Ropework:

- ▪ Use pieces of rope about 15 feet long, depending on the diameter of your spar. Although each spar or pole is nominally 2 inches in diameter, there will be a bit of variation and this may require some adjustment in the length of your ropes.
- ▪ Attach a small piece of duct tape to each end of the rope pieces to prevent fraying.
- ▪ Erecting a watchtower requires knowledge of ropes, knots, and materials and, beyond that, requires the ability to apply this knowledge to securely bind together lightweight building materials into a strong, rigid structure.

There are a few knots you'll need to learn. First are some well-known knots that you might already know: the clove hitch and the bowline. Now, if you don't know how to tie these, don't panic. They are easy to learn and once you know how to tie them, you'll find them useful in many ways apart from building the watchtower. Beyond those generally useful knots are the rope work techniques called lashings. There are many, but only three are required to build the Hour-glass Watchtower: the square lash, the diagonal lash, and the tripod lash. Again, these are easy to learn, but they do take practice to become proficient. These instructions and diagrams, adapted from the *Army Survival Manual*, show you how to tie them.

CLOVE HITCH

Every lashing starts out with a clove hitch. This knot is used to attach a rope or line to a pole. After the clove hitch is tied, then the lashing can begin. Like all good knots, the clove hitch is easy to tie and untie (when the tension on the line is removed).

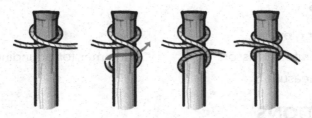

7.11 Clove Hitch

BOWLINE

The bowline is referred to as "the king of knots" because it is so useful. It provides a nonslipping loop in the end of a line that is secure, won't jam up, and is easy to tie and untie.

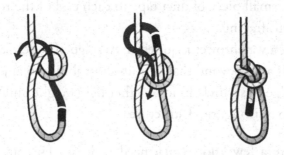

7.12 Bowline

SQUARE LASH

Place two wood or bamboo sticks perpendicular to one another. Make a clove hitch a few inches below the intersection or "transom" on one of the spars. Bring the rope under the transom, up in front of it, horizontally behind the upright, down in front of the transom, and back behind the upright at the level of the bottom of the transom and above the clove hitch.

Each succeeding turn of rope is carefully and tightly wrapped outboard the previous ones on one spar and inboard on the other, not riding over the turns already made. Make the wraps as neat and symmetrical as possible. Four turns or more are required. A couple of "frapping" or tightening turns are then taken between the spars, around the lashing. Tie off the rope with another clove hitch.

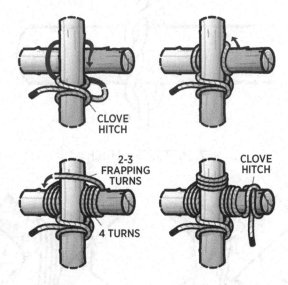

7.13 Square Lashing

7.14 Square Lash

DIAGONAL LASH

Place the two spars perpendicular to one another as before. Tie a timber hitch around both spars and pull tight. Then, wrap four or five turns of rope vertically around the transom and then four or five times around it horizontally. Finish with a few frapping or tightening turns. If necessary, you can jam in a wooden wedge to tighten the connection.

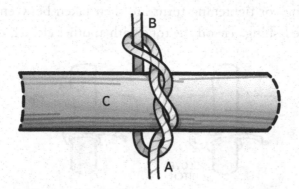

7.15 Timber Hitch

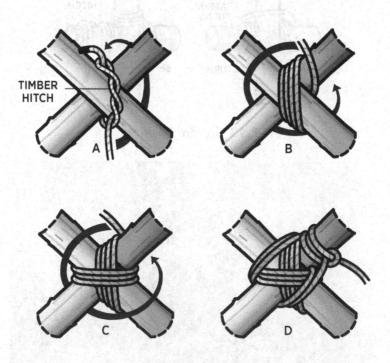

7.16 Diagonal Lash

7.17 Diagonal Lash

TRIPOD LASHING

Start by tying a clove hitch around one pole. Pass the line over and under the three spars as shown six times. These passes are called racking turns. Make the racking turns fairly tight but do leave about ½ inch of space between each spar and its neighbor. Then, make two or three turns as tightly as possible around each racking turn so the racking turns squeeze tightly about each spar. Finish with a clove hitch.

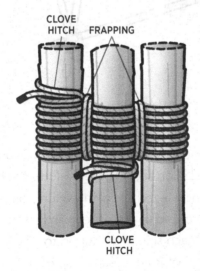

7.18 Tripod Lashing

With these few knots, you'll be able to construct not just the Hourglass Tower, but lashed structures of all sorts!

DIRECTIONS

1. Begin by tripod lashing 3 8-foot Spars together. Position the tripod so that there are exactly 6 linear feet between the points where each spar touches the ground.

2. Next, square lash an 8-foot Pole to each side of the tripod about a foot up from the ground.

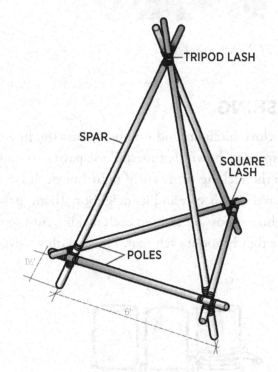

7.19 Lower Tripod

3. Now, loosely tripod lash the three remaining 8-foot Spars together. As you'll soon see, in order to form the hourglass, the tripod must be undone, so this is only a temporary connection for now. Position the tripod so there are exactly 5 linear feet between the points where each side of the tripod touches the ground.

4. Square lash an 8-foot Pole to each side of the tripod you just made about one foot up from the ground. Make these square lashes strong and secure. You can use a saw to cut away some of the overhang if you want, although it's not really necessary.

5. Square lash a 4-foot Platform Rail to each side of the 5-foot-base tripod you just made about about halfway between the apex and bottom end of the Spar. Make these square lashes strong and secure.

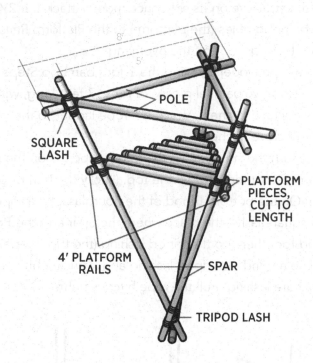

7.20 Upper Tripod

6. Now comes the part that's a bit tricky. Turn both tripods so they rest on their sides. Untie the loosely tied tripod lashing and then weave the Spars through the other tripod so that each one of the un–tripod-lashed Spars contacts one of the tripod-lashed ones. It should look like an hourglass tipped on its side. Now, retie the apex of the three untied Spars together using a tripod lashing. It's a bit harder to do this tripod lashing knot than the previous ones because the three Spars do not lie flat anymore, but it is not terribly difficult. Pull the frapping turns as tight as possible so the connection is tight and does not slip.

7. Turn the structure so that the 6-foot tripod is on the bottom and the 5-foot structure on the top. Have a helper hold the upper tripod in place while you diagonally lash the 3 points where the Spars from one tripod touch the Spars from the other

tripod. Eyeball the structure to make certain the hourglass is straight and not crooked. When the structure is true, frap the diagonal lashes as tightly as possible and tie off. Your tower is beginning to take shape!

8. Turn the structure on its side once more. Attach the 2½-feet-long bamboo poles (the platform) to the Platform Rails with square lashings. Cut off the overhang.

9. Make a rope ladder by tying the 1-foot bamboo Steps at 9-inch intervals to two parallel pieces of rope, 7 feet long, with a series of clove hitches. Attach the rope ladder to the Platform Rail with clove hitches.

10. Again, turn the structure so the 6-foot tripod is on the bottom and the 5-foot tripod is on the top. (Lucky the bamboo is so light!) On the bottom tripod of the hourglass, tie a rope from each square lash—the place where the Spar and the Pole are bound together—to the tripod lashing (the tripod apex). Use a bowline around the tripod lashing and a trucker hitch around the square lashing. Pull the rope hitches tight.

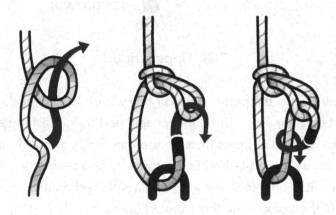

7.21 Trucker Hitch

Alternatively, you can keep guy lines tight by using a "Spanish windlass." Instead of a single rope, use a rope loop for your guy lines. Insert a stick through the loop near the bottom, between the stake and the tower connection, and start twisting. As you twist, the loop pulls together, shortening it and pulling the guy line tight. When the guy line is tight enough, slide the stick almost all the way through the loop and let the long end dig into the ground or catch the legs to hold it in place.

7.22 Spanish Windlass

11. On the top tripod of the hourglass, repeat what you just did on the bottom: tie a rope around each square lash junction to the tripod lashing. Again, pull the ropes tight or use a Spanish windlass.

12. Run a line from the three platform rail square lashings to a stake in the ground, to provide insurance against tipping. Staking the structure provides a great deal of additional stability and is required to safeguard against tipping. The 3-2-1 staking method shown in **diagram 7.23** shows how to prevent the stakes from working loose under load. Drive three rows of large stakes into the ground in the line of pull. The head of each

stake except the last is secured by lashing it to the foot of the stake next behind it. Make a small Spanish windlass by inserting a smaller stick between the strands of rope lashing one stake to another and turning it until there is a fair amount of tension in the rope. Then, drive the point of the smaller stick into the ground to hold it in place.

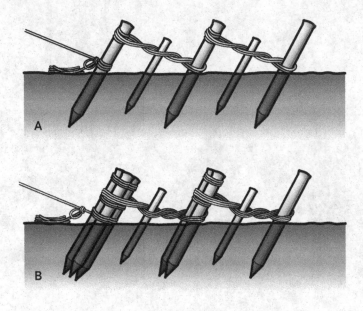

7.23 3-2-1 Staking Method

By now, you've become expert in constructing a variety of defenses, some simple and practical, and some, I admit, preposterously elaborate. It's better than even money that if your house comes anywhere close to actually resembling the fortress of **diagrams 7.24**, your neighbors will leave you alone—completely alone. Your home, once Viking-, Hun-, Mongol-, and Macedonian-proofed through the addition of moats, palisade walls, and watchtowers, will be darn near unconquerable, at least if the problem involves a tribe of preliterate, pre-gunpowder bullies.

But as we'll discuss in the conclusion of this book, the point to all this discussion about palisade walls and steel-jacketed doors isn't really about building an invader-proof home. It is about science, technology, and history.

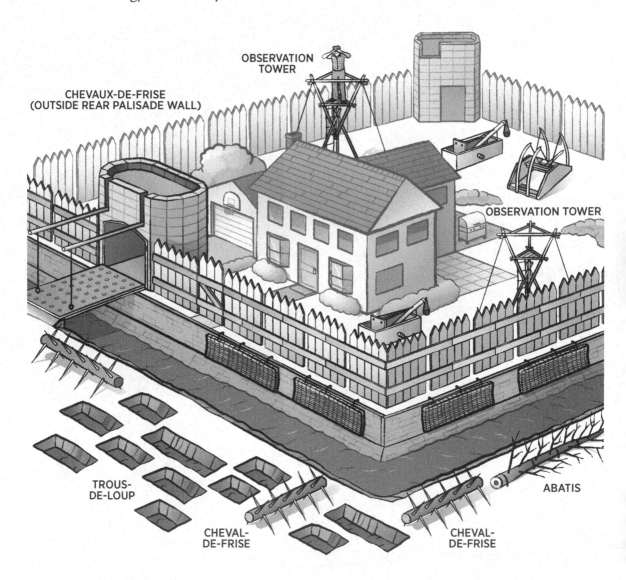

7.24 This fully defended home includes trous-de-loup, observation towers, cheval-de-frise, and other improvements, making it nearly impervious to historical invaders of all types.

8

CONCLUSION

I am sure you'll agree that in these pages we've met a number of unpleasant, nasty groups and their diabolical leaders. Happily, the chances of having to face a Hun or Mongol attack are pretty low.

There are significant differences between these groups. So who takes the prize for the baddest of the bad, the worst of the worst? This summary table shows one man's opinion (mine) as to who you'd least like to see storming your castle.

Vikings The only invaders popular enough to have a professional sports team named after them, the Vikings were sort of mobster-like in that they could be paid off and had funny names. Sure, they were pretty unlikable if it happened to be your monastery or village they were shaking down for money, and they did plunder and pillage. Still, from the historical distance of 1,200 years later, they don't seem quite that bad.

Huns About all we know about them is their name, which is pretty scary. Well, we know a bit more than that: they were a formidable fighting force with an awful reputation, but the details on real atrocities are scarce. Odds are they were dogged, fearless fighters with a mean streak and a penchant for playing very rough.

Macedonians Equipped with great-looking uniforms, they never lost a battle. Tough guys, but not completely unreasonable. A lot of former opponents joined up with them after losing. In fact, toward the end of Alexander's campaign, most of the Macedonian army were actually Persians. On the other hand, they massacred civilians and did a considerable amount of enslaving.

The Peasants' Crusaders Look up *bigoted* and *intolerant* in the dictionary, and you'll see their picture. Besides that, they were loutish, brutal, and none too bright (remember, some of them chose an enchanted goose as their leader). Most of the Peasants' Crusade's human leaders were intriguing, deceitful, and avaricious. All in all, this gang of murderous, rampaging berserkers is not a group to spend much time with if you can avoid it.

Mongols under the Great Khans Scary bad actors; they smelled bad and were doggedly determined to do damage. The Mongols were the remorseless killers of around 20 million people. Massacres of civilians were commonplace, and they treated the people they captured like human shields or beasts of burden. Enslavers and mass murderers on a scale that can scarcely be contemplated, the Mongols were the most pitiless and inhuman group ever known. If it weren't for the next group, that is.

Tamerlane's Tatars Off the scale in terms of evil and cruelty. Tamerlane and his Tatar henchmen actually seemed to take delight in the mass murder of innocent civilians. With an unequaled ferocity, the heartless horsemen from Central Asia took the lives of an estimated 12 million people. This was not quite so high a body count as that amassed by Genghis and his sons and grandsons, but Tamerlane and crew killed nearly as many in a far shorter time.

There were an awful lot of bad actors in the past. And in some places and at some times, there still are. But a little preparation goes a long way. Hopefully, now equipped with the additional knowledge of history and science presented here, you and your loved ones will have far less reason to fear a midnight raid by mounted riders, a charging phalanx of hoplites brandishing sharpened spears, or a stealth attack by a thug with a crowbar out to steal your valuables.

But, as you've very likely figured out by now, the point of this book isn't really to build fortress walls and ancient siege weapons, as much fun as that may be. The real point is to explore the lessons of factual history through the lens of counterfactual science.

The long and glorious relationship between science, technology, and history is worth examining. If you know a bit about it, then you have perspective with which to understand why people and societies behave the way we do. And with that knowledge you can better understand how societies change and how the society you presently live in came to be. When you pick up a newspaper, read the news on the Internet, or even simply look at the way political lines are drawn on the world map, the echoes of Tamerlane, Peter the Hermit, and Genghis Khan still reverberate.

Science and history, taken together, reveal who we really are. American writer and social critic James Baldwin once wrote, "Know from whence you came. If you know whence you came, there are absolutely no limitations to where you can go." If the science and technology is pared away from modern life, the biographies and enumerations of personal experiences of the people who came before us are not unlike our own. So, by looking at the ways that Alexander, Leonardo, or Attila lived and thought, can we not attain a better understanding of ourselves? I think that is more than probable. And there's no better avenue toward such an understanding than the 4,000-year history of how people have learned to defend their castles.

8.1 Defending Your Castle

NOTES

CHAPTER 1: A CAST OF BAD ACTORS

Although we are currently experiencing the longest sustained peace   Steven Pinker, *The Better Nature of Our Angels*, 249.

The only Jerichonian to survive   Joshua 6: 1–27, King James version.

Using scaling ladders, bows and arrows, and other implements of warfare   James Henry Breasted, *Ancient Records of Egypt* Part III, §359; G. Maspero, *History of Egypt, Chaldea, Syria, Babylonia, and Assyria*.

On the night of April 24, 1184 BCE   Joel Munsell, *The Everyday Book of History and Chronology*, 164.

Timur or, as he is better known, Tamerlane   Justin Marzotti, *Tamerlane Sword of Islam*, xxiii.

CHAPTER 2: GENGHIS KHAN AND THE MONGOLS

At the Battle of Liegnitz   Edward Gibbon, *The History of the Decline and Fall of the Roman Empire*, Volume 7.

The conscripted workers were unable to run   Ata-Malik Juvaini, *Genghis Khan: The History of the World Conqueror*.

With this, the Mongols lifted the siege   Stephen R. Turnbull, *Genghis Khan and the Mongol Conquests, 1190–1400*; Nakaba Yamada, *Ghenkō: The Mongol Invasion* of Japan, 54.

"Whoever wishes to fight against the Mongols . . ."   *The Mongol Mission: Narratives and Letters of the Franciscan Missionaries in Mongolia and China in the Thirteenth and Fourteenth Centuries*; *The Long and Wonderful Voyage of Frier John de Plano Carpini*, chapter 18.

As Carpini surmised, even the Mongol horsemen   Vic Hurley, *Arrows Against Steel: The History of the Bow and How It Forever Changed Warfare*, 198.

The Mongols suffered a rare defeat   "Mongol Conquests of the 13th Century," *The Great Soviet Encyclopedia* (1979).

At the Battle of Liegnitz in 1241   Turnbull, *Genghis Khan*, quoting *The Annals of Jan Dlugosz*, 52.

". . . huge lance with a giant X painted on it"   www.allempires.com/article/index.php?q=battle_liegnitz.

After he was killed in battle, his head was removed   Erik Hildinger, *Warriors of the Steppe: Military History of Central Asia, 500 BC to 1700 AD*, 144.

CHAPTER 3: ATTILA AND THE HUNS

During the retreat from Orleans a Christian hermit Friedrich Schiller and Sir Edward Shepherd Creasy, *The World's Cyclopedia of History.*

In just 60 days the walls were rebuilt Neslihan Asutay-Effenberger, *Die Landmauer von Konstantinopel-Istanbul.*

Numerous archaeological excavations have revealed Stijn Arnoldussen, *A Living Landscape: Bronze Age Settlement Sites in the Dutch River Area.*

CHAPTER 4: ALEXANDER THE GREAT AND THE GREEKS

The Greeks fought like lions Jacob Abbott, *Histories of Cyrus the Great and Alexander the Great*, 160.

CHAPTER 5: TAMERLANE AND THE TATARS

But after Melzi died, the notebooks were given away or sold Museo Nazionale Della Scienza e Della Technologia Leonardo Da Vinci, www.museoscienza.org/english/leonardo/manoscritti.

Doubtlessly, he kept his best ideas with him Paul Strathern, *The Artist, the Philosopher, and the Warrior*, 222.

CHAPTER 6: PETER THE HERMIT AND HIS CRUSADERS

"It is indeed the will of God," William Jones, *The History of the Christian Church from the Birth of Christ to the XVIII Century*, v. 1–2.

The goose, however, soon died Norman Roth, *Medieval Jewish Civilization: An Encyclopedia*, 240.

Thousands of defenseless men, women August C. Krey, *The First Crusade: The Accounts of Eyewitnesses and Participants*, 54–56.

Radiocarbon dating estimates W. Neubauer, M. Doneus, A. Eder-Hinterleitner, and K. Löcker, "The Early Neolithic Monumental Enclosure Weinsteig-Grossrussbach," *Archaeologia Polona* 41: 236–238 (2003).

CHAPTER 7: RAGNAR LODBROK AND THE VIKINGS

One of the most famous Viking raiders Martina Sprague, *Norse Warfare: Unconventional Battle Strategies*, 229.

Siegfried's Vikings were a force of 30,000 men Text in Bouquet, *Recueil des Historiens des Gaules et de las France*, Vol. VIII, 4–26; www.fordham.edu/halsall/source/843bertin.asp.

In response, the French built their own catapults Jim Bradbury, *The Routledge Companion to Medieval Warfare*, 133.

the Vikings gave up and simply dragged their longboats Sprague, *Norse Warfare.*

The famous Civil War Union general William Tecumseh Sherman William T. Sherman, *Memoirs of Gen. W. T. Sherman, 1891*, vol 2, 394–397.

These include the trous-de-loup (pits), abatis Dennis Hart Mahan, "A Treatise on Field Fortification" (1852).

SELECTED BIBLIOGRAPHY

Abbott, Jacob. *Histories of Cyrus the Great and Alexander the Great* (New York: Harper & Brothers, 1880).

Bradbury, Jim. *The Routledge Companion to Medieval Warfare* (New York: Routledge, 2004).

Breasted, James Henry. *Ancient Records of Egypt,* Part III.

Gibbon, Edward. *The History of the Decline and Fall of the Roman Empire*, Volume 7.

Hildinger, Erik. *Warriors of the Steppe: A Military History of Central Asia, 500 BC to 1700 AD* (New York: Sarpedon, 1997).

Hurley, Vic. *Arrows Against Steel: The History of the Bow and How It Forever Changed Warfare* (New York: Mason & Charter, 1975).

Jones, William. *The History of the Christian Church from the Birth of Christ to the XVIII Century*, v. 1–2 (Gallatin, TN: Church History Archives, 1983).

Juvaini, Ata-Malik. *Genghis Khan: The History of the World Conqueror*, trans. by John Andrew Boyle (Seattle: University of Washington Press, 1997).

Krey, August C. *The First Crusade: The Accounts of Eyewitnesses and Participants* (Princeton: 1921).

The Long and Wonderful Voyage of Frier John de Plano Carpini (ebooks@adelaide).

Mahan, Dennis Hart. "A Treatise on Field Fortification" (1852).

Marzotti, Justin. *Tamerlane Sword of Islam* (Cambridge, MA: Da Capo Press, 2006).

Maspero, G. *History of Egypt, Chaldea, Syria, Babylonia, and Assyria* (London: Grolier, 1903).

The Mongol Mission: Narratives and Letters of the Franciscan Missionaries in Mongolia and China in the Thirteenth and Fourteenth Centuries (New York: Sheed and Ward, 1955).

Munsell, Joel. *The Everyday Book of History and Chronology* (New York: Appleton, 1858).

Pinker, Steven. *The Better Nature of Our Angels* (New York: Viking, 2011).

Roth, Norman. *Medieval Jewish Civilization: An Encyclopedia* (New York: Routledge, 2003).

Schiller, Friedrich, and Sir Edward Shepherd Creasy, *The World's Cyclopedia of History* (New York: Alden, 1883).

Sprague, Martina. *Norse Warfare: Unconventional Battle Strategies* (New York: Hippocrene Books, 2007).

Strathern, Paul. *The Artist, the Philosopher, and the Warrior.* (London: Vintage, 2010).

Turnbull, Stephen R. *Genghis Khan and the Mongol Conquests, 1190–1400* (New York: Routledge, 2004).

US Army Survival Manual FM 3-05.70, May 2002.

Yamada, Nakaba. *Ghenkō: The Mongol Invasion of Japan.* (Washington, DC: University Publications of America, 1979).

INDEX